Food Colloids: 3rd Edition

Food Colloids: 3rd Edition

Guest Editors

Eleni P. Kalogianni
Julia Maldonado-Valderrama

MDPI

Basel • Beijing • Wuhan • Barcelona • Belgrade • Novi Sad • Cluj • Manchester

Guest Editors

Eleni P. Kalogianni
Department of Food Science and Technology
International Hellenic University
Thessaloniki
Greece

Julia Maldonado-Valderrama
Department of Applied Physics
University of Granada
Granada
Spain

Editorial Office
MDPI AG
Grosspeteranlage 5
4052 Basel, Switzerland

This is a reprint of the Special Issue, published open access by the journal *Colloids and Interfaces* (ISSN 2504-5377), freely accessible at: https://www.mdpi.com/journal/colloids/special_issues/D33RDVGGN0.

For citation purposes, cite each article independently as indicated on the article page online and as indicated below:

Lastname, A.A.; Lastname, B.B. Article Title. *Journal Name* **Year**, *Volume Number*, Page Range.

ISBN 978-3-7258-7210-7 (Hbk)
ISBN 978-3-7258-7211-4 (PDF)
https://doi.org/10.3390/books978-3-7258-7211-4

Contents

About the Editors

Eleni P. Kalogianni

Eleni P. Kalogianni is a Professor in the Department of Food Science and Technology, School of Geosciences, International Hellenic University (I.H.U.), Greece. She is Vice President of the European Federation for the Science and Technology of Lipids and a national representative for the same organization. She currently heads the Oils and Interfaces research group and specializes in food science and technology, with a specific focus on the study of food colloids and interfacial properties, as well as on the extraction, processing, and characterization of lipids. She has research experience lasting over 20 years, she has published more than 65 JCR-indexed articles and has been the coordinator/head of a series of competitive research projects funded by Greek, European, and private grants. She has organized and chaired scientific conferences and one-day seminars and, in April 2024, the 19th Food Colloids Conference. She was head of the Food Science and Technology department for 2021–2023 and has also been a member of the I.H.U. Research Committee. She represents I.H.U. on the Greek Technological Platform "Food for Life" (a member of the European Platform) and coordinates the working group "Sustainable food production" under the aegis of the Federation of Hellenic Food Industries.

Julia Maldonado Valderrama

Julia Maldonado Valderrama holds a PhD in Science and Technology of Colloids and Interfaces and a Degree in Physics from the University of Granada, Spain. She has worked in international laboratories including the Max Planck Institute of Colloids and Interfaces (Germany), the Laboratorie de Physique des Solides (France), and the Institute of Food Research (UK) as a Marie-Curie Fellow, and more recently worked as a visiting professor at the University of South Australia (Australia). She is currently Associate Professor at the Department of Applied Physics, UGR, and has three children born in 2012, 2014 and 2016. Her research lines focus on the interfacial characterization of foams and emulsions, in vitro digestion, and the development of biomedical and food nanosystems, having coordinated national and international research projects and one contract with industry (Kemin). She has published more than 75 JCR-indexed articles resulting in more than 3000 citations and she is involved in academic mentorship, having supervised PhD students, master's and final degree projects, and more than 20 international visitors at her laboratory. She teaches undergraduate and postgraduate subjects and is currently Vice Dean of Students and Employability at the Faculty of Sciences, UGR. She is also member of the executive board of the Specialized Group of Colloids and Interfaces, as well as national and international advisory committees. She is the Chair of the 20th Food Colloids conference in 2026 and regularly engages in science and public outreach activities.

colloids and interfaces

MDPI

Editorial

Special Issue "Food Colloids: 3rd Edition"

Eleni P. Kalogianni [1,*] and Julia Maldonado-Valderrama [2,*]

1 Oils and Interfaces Engineering Group, Department of Food Science and Technology, School of Geotechnical Sciences, International Hellenic University, 57400 Thermi, Greece
2 Biocolloid and Fluid Physics Group, Department of Applied Physics, University of Granada, 18071 Granada, Spain
* Correspondence: elekalo@ihu.gr (E.P.K.); julia@ugr.es (J.M.-V.)

This Special Issue features a collection of research papers following the 19th Food Colloids Conference, organized by the International Hellenic University and held on the 14–18 April 2024 in Thessaloniki, Greece. This international Conference focused on the theme "Using Colloid Science to Find New Sustainable Solutions in Food," reflecting the growing importance of colloid science in addressing global challenges related to food sustainability, nutrition, and health. Nowadays, food colloids comprise a multidisciplinary platform where chemistry, physics, biology, and engineering converge to design functional food systems. The rational design of these systems relies on a deep comprehension of the molecular, physicochemical, interfacial, and colloidal properties of food ingredients, and how they are digested and delivered. This Special Issue showcases some of the latest advances in colloidal structures including emulsions, foams, gels, and nanoparticles, with applications ranging from nutrient delivery and food preservation to safety and sensory enhancement.

In the field of **Bioactive Encapsulation and Delivery**, new ingredients with bioactive properties are encapsulated using different matrices. Madybekova et al. used alginate-chitosan to efficiently encapsulate *Bifidobacterium bifidum* and Stevia leaf extract. This system, stabilized by strong polymer interactions verified by FTIR, targets the enrichment of a nutrient-rich acid whey beverage. Vaishnav et al. expslore carrageenan encapsulation of snail protein hydrolysates importantly enhancing their stability and bioactivity. Although immediate antioxidant activity and solubility were marginally reduced, the structural protection provided suggests suitability for controlled-release nutraceutical applications. Chitosan was also applied in combination with tripolyphosphate to produce nanoparticles to encapsulate ethanolic cashew leaf extract, as a source of polyphenols by Sinapapanya et al. The resulting formulation demonstrates strong antimicrobial and antioxidant activities importantly supporting the use of plant-based bioactives in food preservation.

Two contributions deal with **Nanoemulsion** Stability and Safety Assessment. On one hand Silva-Paz et al. develop optimized nanoemulsions of canola oil stabilized with soy lecithin and Tween 80 introducing protein gels to modify textural properties. Their rheological analysis suggests promising applications in food texture and stability enhancement. On the other hand, Hendawy et al. explore the safety of fish byproducts/lemon oil using nanoemulsions in aquatic models. Their findings highlight the need for balanced formulation strategies to ensure both efficacy and safety in food and pharmaceutical applications.

Anh et al. demonstrate the new advanced functionalization of **colloidal silver nanoplates for food safety monitoring.** The nanoplates, synthesized at room temperature, provided very high detection sensitivity as evidenced by an ultralow limit of detection for a critical antibiotic used in livestock, innovating in food safety monitoring.

New sustainable materials are evaluated in the final contributions, in which microalgae biomass is explored as a sustainable foaming agent by Georgiou et al. The protein-rich biomass extract displays a high foaming capacity and stability, suggesting high potential for commercial use in instant non-milk foam-forming beverages. Finally, Sharma et al. review the application of sodium alginate, a green, non-toxic biopolymer derived from seaweed, as an ideal source for edible fruit coatings. Its film-forming capacity and ability to retain moisture makes it a promising agent to develop coatings for shelf-life extension. Plant extracts or nanoparticles can be added to overcome their inherent hydrophilicity, also achieving enhanced antimicrobial and antioxidant effects.

The contributions in this Special Issue reflect the dynamic and interdisciplinary nature of food colloid science. From advanced encapsulation techniques to novel applications in food safety and functionality, the research presented here underscores the critical role of colloids in developing sustainable, health-promoting food systems. The Food Colloids Conference series reinforces importance of colloid science, inspiring further research and industrial applications and shaping the future of food.

Author Contributions: Conceptualization J.M.-V. and E.P.K.; writing—original draft preparation, J.M.-V. and E.P.K.; writing—review and editing, E.P.K. All authors have read and agreed to the published version of the manuscript.

Conflicts of Interest: The authors declare no conflict of interest.

List of Contributions:

1. Madybekova, G.; Turkeyeva, E.; Mutaliyeva, B.; Osmanova, D.; Aidarova, S.; Miller, R.; Sharipova, A.; Issayeva, A. Study of Probiotic Bacteria Encapsulation for Potential Application in Enrichment of Fermented Beverage. *Colloids Interfaces* **2024**, *8*, 51. https://doi.org/10.3390/colloids8050051.
2. Sinlapapanya, P.; Buatong, J.; Palamae, S.; Nazeer, R.; Zhang, B.; Prodpran, T.; Benjakul, S. Ethanolic Cashew Leaf Extract Encapsulated in Tripolyphosphate–Chitosan Complexes: Characterization, Antimicrobial, and Antioxidant Activities. *Colloids Interfaces* **2024**, *8*, 52. https://doi.org/10.3390/colloids8050052.
3. Hendawy, A.; Elgharbawy, A.; Mohd Noor, N.; Al-Saari, N.; Azmi, N.; Mohd Salleh, H. Exploring the Safety of the Sustainable Toxicity Testing in Zebrafish and Brine Shrimp Using Nanoemulsions Formulated from Fish Byproducts and Lemon Oil. *Colloids Interfaces* **2024**, *8*, 59. https://doi.org/10.3390/colloids8060059.
4. Georgiou, D.; Charisis, A.; Theocharidou, A.; Ritzoulis, C.; Papapanagiotou, G.; Samara, C.; Chatzidoukas, C.; Kalogianni, E. Foaming Properties of *Chlorella sorokiniana* Microalgal Biomass. *Colloids Interfaces* **2024**, *8*, 66. https://doi.org/10.3390/colloids8060066.
5. Silva-Paz, R.; Ñope-Quito, C.; Rivera-Ashqui, T.; Jamanca-Gonzales, N.; Eccoña-Sota, A.; Riquelme, N.; Arancibia, C. Encapsulation of Canola Oil by Sonication for the Development of Protein and Starch Systems: Physical Characteristics and Rheological Properties. *Colloids Interfaces* **2025**, *9*, 10. https://doi.org/10.3390/colloids9010010.
6. Vaishnav, A.; Mehta, N.; Priyadarshini, M.; Singh, S.; Acharya, P.; Biswal, S.; Nath, H.; Hussain, S.; Pal, P.; Lal, J.; et al. Impact of Carrageenan-Based Encapsulation on the Physicochemical, Structural, and Antioxidant Properties of Freshwater Snail (*Bellamya bengalensis*) Protein Hydrolysates. *Colloids Interfaces* **2025**, *9*, 29. https://doi.org/10.3390/colloids9030029.

7. Anh, C.; Cao, D.; Ngan, L. Preparation of Colloidal Silver Triangular Nanoplates and Their Application in SERS Detection of Trace Levels of Antibiotic Enrofloxacin. *Colloids Interfaces* **2025**, *9*, 31. https://doi.org/10.3390/colloids9030031.
8. Sharma, A.; Singh, A. Sodium Alginate: A Green Biopolymer Resource-Based Antimicrobial Edible Coating to Enhance Fruit Shelf-Life: A Review. *Colloids Interfaces* **2025**, *9*, 32. https://doi.org/10.3390/colloids9030032.

colloids and interfaces

MDPI

Article

Study of Probiotic Bacteria Encapsulation for Potential Application in Enrichment of Fermented Beverage

Galiya Madybekova [1], Elmira Turkeyeva [2], Botagoz Mutaliyeva [2,*], Dinara Osmanova [2], Saule Aidarova [3], Reinhard Miller [4], Altynai Sharipova [5] and Assem Issayeva [3]

1 Chemistry Department, O. Zhanibekov South-Kazakhstan Pedagogical University, Shymkent 160012, Kazakhstan; galiya56@list.ru

2 Biotechnology Department, M. Auezov South-Kazakhstan University, Shymkent 160012, Kazakhstan; turkeeva_1980@mail.ru (E.T.); dinaraosmanova0113@gmail.com (D.O.)

3 "One Belt, On Road" Petroleum Engineering Institute, Kazakh-British Technical University, Almaty 050000, Kazakhstan; ainano9999@gmail.com (S.A.); isa-asem@mail.ru (A.I.)

4 Soft Matter Biophysics, Technische Universität Darmstadt, 64289 Darmstadt, Germany; reinhard.miller@pkm.tu-darmstadt.de

5 Mining and Metallurgical Institute, Satbayev University, Almaty 050013, Kazakhstan; a_sharipova85@mail.ru

* Correspondence: botagoz.mutalieva@auezov.edu.kz; Tel.: +7-7011298090

Abstract: The current work is devoted to the development of probiotic microencapsulation systems with the co-encapsulation of a plant extract, which can increase the survival of beneficial bacteria and are suitable for potential applications in the enrichment of fermented beverages based on acid whey. The encapsulation process exhibited a high level of effectiveness, achieving 83.0% for Bifidobacterium (BB), 89.2% for Stevia leaf extract (SE), and 91.3% for their combination (BB + SE). The FTIR analysis verified substantial interactions between the encapsulated agents and the polymer matrix, which enhanced the stability of the microcapsules. The BB + SE microcapsules exhibited reduced swelling and moisture content, indicating a denser structure compared to separately encapsulated BB and SE. Comparison of release kinetics of BB, SE and BB + SE loaded microcapsules showed that the combination of active agents has a quicker initial release, reaching 60% release within the first 2 h, and this value increased to 70% after 4 h. The release kinetics studies demonstrated a controlled release of active substances over 24 h. A morphology analysis shows that the surfaces of the dry microcapsules containing BB, SE, and their combination BB + SE have a porous structure. For encapsulated agents, the size of the capsules produced with BB and SE are smaller than those produced with two components (BB + SE), the sizes of which are between 760 μm and 1.1 mm. Modeling of the behavior of microcapsules in a simulated gastrointestinal tract provides information on swelling and active agents release rates as a function of pH in real biological environments. Thus, the new formulations of microcapsules with microorganisms and plant extracts have great potential for the development of fermented whey-based beverages.

Keywords: acid whey; encapsulation; *Bifidobacterium bifidum*; *Stevia rebaudiana Bertoni*; active agents; release; morphology; size; swelling degree; encapsulation efficiency

1. Introduction

Whey is a byproduct of cheese production; it is rich in carbon and nitrogen and contains high concentrations of lactose, proteins, and minerals [1,2]. There are two types of whey: sweet whey, which remains after enzymatic milk coagulation, and acid whey, which remains after milk coagulation via acidification [3]. As described in [4], acid whey is limited in its application because of its low pH and lactose content, and it mostly remains unused, creating environmental problems in terms of polluting water, soil, etc. [5].

There are many challenges to utilizing acid whey, but, at the same time, it is very attractive due to its richness in micro- and macro-elements. For example, the calcium content of acid whey is two times higher than that of sweet whey, and this increases the value of acidic whey for obtaining fermented beverages and other products.

With added probiotic bacteria, acid whey can be used to obtain fermented beverages with properties beneficial to health [6]. Many of the health benefits of probiotics are associated with their survivability and stability in carrier food and under gastrointestinal conditions.

Fermented beverages are beneficial to the host only when the probiotics are administered in adequate amounts [7]. The fundamental properties of probiotics are resistance to stomach acid and tolerance to bile salts, as these properties allow them to survive acidic stomach conditions and the presence of bile salts in the small intestine during passage through the GIT [8].

However, several harsh conditions threaten their survival rate during processing, storage, and transit through the gastrointestinal tract (GIT), as well as possible interactions with beverage constituents during the incorporation of probiotics into beverages [9].

As noted by the authors of [10], most probiotics cannot survive in large quantities due to the low pH of gastric juice, which limits their effectiveness in most functional foods. Scientists have studied and compared free and capsulated probiotic microorganisms and concluded that probiotics protected by capsules showed the best results for bioavailability and survival in a gastrointestinal simulation [11].

Traditional techniques for maintaining the survival of probiotics during processing and storage are the selection of resistant strains, the inclusion of protective compounds, oxygen control, adaptation to stress, two-stage fermentation, and the addition of prebiotic substances [12]. Other authors [13] had the idea that enriching the alginate matrix used for probiotic encapsulation with whey as a substrate can have an impact on the antioxidant capacity and stability of whey-based beverages. They used whey as a substrate for fermentation to create a functional beverage and to solve environmental problems such as water contamination. This work emphasizes that encapsulation can not only solve the problem of preserving the viability of probiotics but can also increase the antioxidant properties of functional beverages. Zanjani et al. [14] stated that microencapsulation is one of the most modern methods for aiding probiotic survival and that the development of corresponding delivery systems can allow for the preservation and prolongation of probiotic activity [15]. Yao et al. [15] presented a new strategy for the microencapsulation of probiotics, and they highlighted the key mechanisms of their stress-resistant properties. They reviewed the latest in vitro and in vivo models to evaluate the effectiveness of probiotic delivery systems, and they emphasized the importance of having suitable models that can protect probiotics from aggressive environments.

Thus, probiotic microorganisms are encapsulated to ensure their survival during processing and storage and to protect them from the harmful effects of gastric pH and bile salts in order to ensure maximum bioavailability in the host [16]. Not only is encapsulation a solution to the problem of preserving the viability of probiotics, but it can also increase the antioxidant properties of functional beverages. Furthermore, using prebiotics in combination can selectively promote the growth and activity of beneficial bacteria and significantly maintain the survival of probiotics during processing and storage [12].

Currently, the development of delivery systems (in particular, microencapsulation systems) for various probiotic drugs, with the aim of overcoming the above-mentioned challenges [17], is the topic of many studies.

The composition of encapsulating material is the main determinant of microcapsule functional properties [18], and processing variables such as the concentration, characteristics of the core and coating materials, and stability, as well as the functionality and release kinetics of the desired product, can serve as criteria for the selection of matrix polymers to achieve efficient and effective encapsulation [19].

The effect of various encapsulating materials on the stability of probiotic bacteria was studied in [17], and it was found that microcapsules with alginate, xanthan gum, and carrageenan gum significantly improved the vitality of probiotic bacteria when exposed to acidic conditions and bile acid salts.

Alginate is commonly used to encapsulate probiotics because it is a promising system for their delivery [20]. In comparison with biomolecules such as polysaccharides (chitosan, xanthan, dextrin, and carrageenan) and milk proteins, sodium alginate is used more often because of its low costs and easy accessibility; it is also biocompatible and nontoxic [20,21].

Chitosan is the second most abundant biopolymer in the world, following cellulose. It is a versatile biopolymer with unique properties, and it is nontoxic, biodegradable, and biocompatible [22].

Zanjani et al. [14] found that alginate microcapsules can be coated by oppositely charged biopolymers, such as chitosan, leading to a strong interaction between the carboxylic groups of alginate and the amine groups of chitosan, resulting in enhanced stability of the microcapsules.

Moreover, the choice of appropriate methods is important because probiotics are highly sensitive and can lose their viability during encapsulation under different conditions. Mild techniques like spray-drying, extrusion, emulsification, and lyophilization are suitable with materials such as polysaccharides, proteins, and their complexes [23].

The two most commonly employed methods for obtaining microcapsules are emulsion and non-emulsion techniques. The encapsulation of probiotic drugs using emulsions provides stability by utilizing polyelectrolyte complexes that can be sustained in a wide range of external conditions (pH, ionic strength, and temperature). However, at the same time, this method is more complicated and consists of many formulation steps: an oil-in-water emulsion is produced and then mixed with the encapsulated materials. For water-soluble substances, the process involves creating a primary emulsion followed by the formation of a secondary emulsion, where the active ingredients are covered by envelope materials.

The second technique is simpler and based on direct microencapsulation with biopolymers (chitosan and alginate) without an emulsification step [24,25]. This is more suitable for microorganisms because of the absence of organic solvents.

Recently, not only probiotics but also various additives with prebiotic properties, such as Stevia, have been used to develop functional beverages. Several publications have described various reasons for using prebiotics. Combinations of several polysaccharides (alginate, κ-carrageenan, locust bean gum, gellan gum, and xanthan gum) with prebiotics (resistant starch, lactulose, and lactosucrose) and their effects on the encapsulation of the *Lactobacillus casei* 01 strain were studied, and higher survivability and protection [26] were observed. The addition of natural prebiotics (potato starch, *Plantago psyllium*, or inulin) to systems was found to improve the viability of *Lactobacillus casei* Shirota and two species of *Lactobacillus plantarum* (Lp17 and Lp33) [27]. Functional products such as stirred yogurt fortified with the prebiotic xylooligosaccharide and probiotic and synbiotic microcapsules were described by Ismail et al. [28]. The effect of inulin on the metabolic characteristics of probiotic yeast was studied, and it was found to affect the functionality of ice cream, increasing survivability and allowing for a long storage time [29]. Thus, prebiotics increase the metabolism of probiotics and maintain their survival during processing and storage [30].

The encapsulation of probiotics with Stevia extract as a prebiotic can solve problems related not only to the vitality and stability of beneficial microorganisms but also substitute carbohydrates with high glycemic indices, which improve their effects on human health [31].

Thus, the aim of the present work is to study the development of delivery systems for probiotics with a co-encapsulated plant extract that can increase survivability, prolong the positive nutrition effects on the human intestinal microflora, and have potential application for the enrichment of fermented beverages based on acid whey.

The preferred probiotic microorganisms are members of the genus *Bifidobacterium*, which are used in many foods and often added to dairy products.

There are many scientific reports on the description and health-promoting properties of *Bifidumbacterium* [32]. It is known that *Bifidumbacterium* (belonging to *the* phylum Actinobacteria) is the most commonly used and safe human probiotic, as certified by the

Food and Drug Administration (FDA) with the acronym GRAS (Generally Recognized As Safe) and by the European Food Safety Authority (EFSA) with the acronym QPS (Qualified Presumption of Safety); they are also acid-tolerant bacteria [33,34]. From this point of view, they can be used for the enrichment of beverages based on acid whey.

The study hypothesis is that delivery systems loaded with a probiotic culture and a natural plant sweetener with prebiotic properties have potential application for the enrichment of fermented whey beverages and can impart beneficial properties to the final product.

2. Materials and Methods

2.1. Materials

Chitosan (5–20 mPa·s, 0.5% in 0.5% Acetic Acid at 20 °C), manufacturer Tokyo Chemical Industry Co., Ltd. (TCI), Tokyo, Japan.

Calcium chloride dihydrate, certified ACS, Thermo Fisher Scientific Inc., Waltham, MA, USA. Sigma-Aldrich (Merck, Darmstadt, Germany) supplied Sodium alginate powder (CAS Number: 9005-38-3, Viscosity 15–25 cps).

Bifidobacterium bifidum 791, which was proposed as a starter culture by Voblikova [35], was used as a probiotic microorganism, and the plant extract of *Stevia rebaudiana Bertoni* was used as an agent for co-encapsulation. These materials were provided by the laboratory "Biotechnology" of the M. Auezov South Kazakhstan University. To isolate bifidumbacteria, selective MRS (69964-500G, MRS Agar, suitable for microbiology, made in Switzerland, sigmaldrich.com) medium of a modified composition was used by adding cysteine (Sigma-Aldrich (Merck)), which is necessary for their growth, in anaerobic conditions. The release of the microencapsulated agent was studied in a phosphate-buffered solution (PBS). Standard PBS was prepared from sodium chloride, potassium chloride, potassium dihydrogen phosphate, and disodium hydrogen phosphate (Sigma-Aldrich (Merck)). All reagents were analytical grade, and used without additional purification.

2.2. Methods

2.2.1. Microcapsule Preparation

Microcapsules were prepared in a two-stage process via ion gelation and the complexation of polyelectrolytes at an ambient temperature, as described in [25]. *Bifidumbacterium* at a concentration of 1×10^8 CFU/g was added to 100 mL of sodium alginate solution (1.5%) and homogenized via gentle stirring with a magnetic stirrer for 60 min. Then, the suspension was dropped via a syringe pump (NE-1002X Programmable Microfluidics, New Era Pump Systems, Inc., Farmingdale, NY, USA) into 100 mL of 2% $CaCl_2$ solution as a cross-linking agent, under constant magnetic stirring, and the resulting microcapsules were washed with distilled water and filtered through a Buchner funnel (Coors™ Buchner funnel, Merck, Darmstadt, Germany).

In the second stage, the washed formulations were dispersed in 50 mL of chitosan solution (0.5% CS in 0.5% CH_3COOH (Sigma-Aldrich, Merck)) under constant stirring (magnetic stirrer). The chitosan in the formulations of the polymeric capsules provided the formation of polyelectrolyte complex shells of two polymers; this can stabilize the structure of the capsules, which have properties that maintain the integrity in an acidic environment simulating the stomach and dissolve in a medium simulating the pH of the intestines [36].

The microcapsules were prepared with different active agents: (CS/(ALG/(Ca + BB))); (CS/(ALG/(Ca + SE))); (CS/(ALG/(Ca + BB + SE))).

The release profile of the active agents from the microcapsules was studied in vitro by UV-Vis spectrophotometry ((UV-1900i dual-beam spectophotometer with software (190–1100 nm), scanning speed up to 29,000 nm/m, manufacturer Shimadzu, Japan).

2.2.2. Determination of the Encapsulation Efficiency and Loading Capacity

The encapsulation efficiency (EE) was determined spectrophotometrically from the total concentration of active agents (c_{total}) and the content in the dried microcapsules (c_{load})

according to a method described by Vlahoviček-Kahlina et al. [24]. The encapsulation efficiency was calculated using the equations in [24]:

$$EE = (c_{load}/c_{total}) \times 100$$

$$c_{load} = c_{total} - c_f,$$

where c_f is the concentration of the active agent in the filtrate.

To determine the EE, the solution was stirred and left in the dark for 15 min, and the value on the spectrophotometer was read for samples with BB, SE, and BB + SE at λ = 595 nm.

The active agent contents (BB, BB + SE, and SE) in the microcapsule LC were determined by using a method described by Vlahoviček-Kahlina et al. [24]:

$$LC = w_{active\ agent}/w_c$$

where $w_{active\ agent}$ is the weight of loaded active agent and w_c is the weight of the microcapsules/beads.

2.2.3. Determination of the Degree of Swelling

The swelling process depends on the property of the dissolving agent, among other things. Therefore, to avoid the influence of the electrolytes in the buffer solutions, the degree of swelling (S_w) was determined for microcapsules dispersed in distilled water, as described by Kudasova et al. [25]. The microcapsules (10 mg) were transferred to test tubes, and they were allowed to swell at room temperature for three hours in 10 mL of distilled water to reach equilibrium. The weight of the wet swollen microcapsules was determined after removing the moisture from their surfaces with filter paper. S_w was calculated using a formula proposed by Vlahoviček-Kahlina et al. [24]:

$$S_w/\% = ((w_t - w_0/w_0))$$

where w_t is the weight of swollen microcapsules and w_0 is the weight of the dry capsules.

All measurements were repeated three times, and the results are presented as the mean value with the corresponding standard deviation.

2.2.4. Fourier Transform Infrared Spectroscopy Analysis

Fourier transform infrared spectroscopy spectra were recorded with an infrared spectrometer, a Nicolet 5700 "Thermo Electron Corporation", Waltham, MA, USA, powered by OMNIC™ v9.x software, with the spectral field in the range of 500–4000 cm^{-1}.

2.2.5. Determination of Morphology

Analyses of the morphology, size, and shape of the wet and dry microcapsules were carried out with an optical microscope (Olympus Soft Imaging Solutions GmbH, version E_LCmicro_09Okt2009, Tokyo, Japan) immediately after their preparation and with a scanning electron microscope (SEM) (JEOL-SEM, JSM-6490 LV (Jeol Ltd., Tokyo, Japan), with INCA Energy 350 energy X-ray microanalysis (Oxford instruments, Oxford, UK) and HKL Basic (Oxford instruments, UK) and an SEM Quanta 3D 200i (Thermo Fisher Scientific, USA), after drying to a constant mass.

Scanning electron microscopy (Quanta 3D 200i) was carried out on all samples at an accelerating voltage of 15 keV and a working distance of ≈12 mm. All samples were secured with double-sided graphite 9 tape. All measurements were carried out in the high-vacuum mode at 10^{-3} Pa. The maximum magnification was ×5000, and the minimum magnification was ×40.

2.2.6. Release of Active Agents from the Microcapsules

The release profile of the active agents from the microcapsules was studied spectrophotometrically in vitro. Absorption spectra were measured at 595 nm to avoid damage to bacterial cells. All measurements were performed in phosphate-buffered solutions (pH 7.4).

In general, this study indicates that microencapsulation with an alginate-gelatinized starch coated with chitosan could successfully and significantly protect probiotic bacteria from adverse conditions simulating the human gastrointestinal environment.

2.2.7. Modeling the Behavior of Microcapsules with Active Agents in an Imitating Gastrointestinal Tract

The release kinetics of microcapsules in the imitating GIT have been modeled using a modified Higuchi model [37].

In this research, a linear model had been used based on the interpolation of selected data, which provides a precise approximation.

For a given time t, if $t_1 \leq t \leq t_2$, the release percentage $R_{(t)}$ is calculated as:

$$R_{(t)} = R_1 = \frac{R_2 - R_1}{t_2 - t_1} \times (t - t_1)$$

where t_1 and t_2 are the consecutive time points, R_1 is release percentage at t_1, R_2 is the release percentage at t_2, t is the studied time, and R_t is the calculated release percentage at time t.

The swelling behavior of the active agent loaded microcapsules have been modeled using an exponential function in dependence of time and pH of GIT [38]:

$$S_t = S_0 \times \left(1 + k \times \left(1 - e^{\alpha t}\right)\right.$$

where S_0 = 1—initial swelling factor, k is a constant related to the swelling capacity in dependence of GIT pH, and α is a rate constant (in the dependence of GIT pH), which shows how quickly the swelling occurs.

3. Results and Discussion

The studied systems were composed of sodium alginate (dry matter); calcium chloride; distilled water; Stevia powder; *Bifidumbacterium*; and the encapsulated agents (1) BB, (2) SE, and (3) BB + SE.

The physicochemical characteristics of microcapsules obtained according to the modified method [25] were studied.

3.1. Encapsulation Efficiency, Loading Capacity and Swelling Degree of Microcapsules

The loading efficiency and loading capacity were determined to obtain information on the yield and content of the active agents in the chitosan/alginate microcapsules. It is known that the entrapped amounts of active agents in chitosan/alginate microcapsules depend on the type and concentration of the gelling cation and active agents, as well as the preparation method [39].

The significant differences in the encapsulation efficiency (EE) and swelling degree (S_w) between the samples are shown in Tables 1 and 2. The difference in EE reflects the extent of electrostatic interactions and hydrogen bonds involved in the interactions of active agents with both calcium chloride and sodium alginate during the preparation. The most water-soluble active agent, BB + SE, exhibits the highest EE values, indicating a greater extent of molecular interactions in the solution. In addition to the molecular interactions, the encapsulation capacity also depends on the structure of the active agents.

Table 1. Encapsulation efficiency (EE) of the microcapsule formulation CS/(ALG/(Ca + active agent)).

Encapsulated Active Agent	EE_1, %	EE_2, %	EE_3, %	Average Value	Standard Deviation	Mean Square Deviation
BB	80.0	89.0	80.0	83.0	5.20	4.24
SE	89.6	88.09	89.0	89.2	0.38	0.31
BB + SE	90.0	92.0	92.0	91.3	1.15	0.94

Table 2. Swelling degree (S_w) of the microcapsule formulation CS/(ALG/(Ca + active agent)).

Encapsulated Active Agent	S_{w1}, %	S_{w2}, %	S_{w3}, %	Average Value	Standard Deviation	Mean Square Deviation
BB	76.9	80.0	80.0	79.0	1.78	1.45
SE	76.9	71.4	71.4	73.3	3.18	2.59
BB + SE	53.9	55.5	50.0	53.1	2.82	2.30

The encapsulation efficiency results—83.0% for *Bifidumbacterium* (BB), 89.2% for Stevia extract (SE), and an impressive 91.3% for their combination (BB + SE)—highlight the effectiveness of the encapsulation process. These results not only confirm the success of a well-optimized encapsulation method, as supported by previous studies [13,15,17], but they also underscore the innovative potential of co-encapsulation in enhancing the functionality of bioactive compounds.

The highest encapsulation efficiency observed with BB + SE co-encapsulation suggests a synergistic effect, likely resulting from complementary interactions between the probiotic and the plant extract during the encapsulation process. These interactions appear to enhance both retention and stability of the active agents within the microcapsules, as noted by Xu et al. [40]. Additionally, Stevia extract seems to increase the matrix's affinity for bacterial cells, further improving encapsulation efficiency, as highlighted by Yao et al. [15]. This superior encapsulation efficiency is particularly beneficial for functional food applications, where maintaining the stability and controlled release of probiotics is crucial for delivering sustained health benefits [41].

The enhanced encapsulation efficiency is further corroborated by the lower swelling degree observed in the BB + SE microcapsules (53.1%), compared to microcapsules containing only BB (79.0%) or SE (73.3%). This reduced swelling can be attributed to stronger intermolecular interactions and enhanced cross-linking within the microcapsule matrix, as co-encapsulation of multiple active agents often leads to a more compact structure [14]. The lower swelling degree is advantageous for applications requiring controlled release and prolonged stability, as it typically correlates with slower release rates and better retention of encapsulated materials [12,41].

Moreover, the co-encapsulation of BB and SE results in the formation of larger microcapsules with a lower moisture content (approximately 55%), compared to the smaller, more hydrated microcapsules containing only BB (up to 77% moisture). This difference suggests that co-encapsulation promotes a more compact and less hydrated microcapsule structure, likely due to the enhanced interactions between the encapsulated components and the matrix. The larger size and reduced moisture content of the BB + SE microcapsules may contribute to improved stability and controlled release, further supporting their potential effectiveness in delivering active agents in functional food applications.

3.2. FTIR Characterization of Microcapsules Encapsulating BB, SE, and Their Combination BB + SE

Fourier transform infrared spectroscopy (FTIR) provides information for the identification of chemical bonds according to the infrared absorption spectrum. The spectra reveal the profile of a sample containing various components. The spectra of the BB, SE, and their mixtures with core components are presented and discussed in Figure 1.

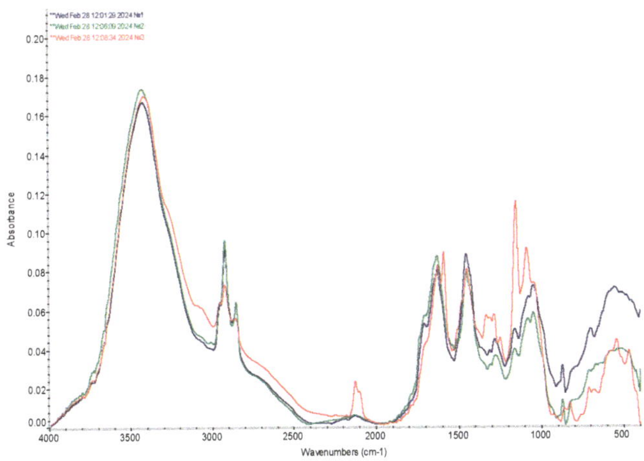

Figure 1. FTIR spectra of microcapsules loaded with *Bifidumbacterium*, Stevia extract, and mixture of *Bifidumbacterium* with Stevia extract. №1 Bifidumbacterium bifidum (BB); №2 Stevia extract (SE); №3 BB + SE.

The FTIR spectra of the microcapsules containing BB exhibit a notable peak near 3413.9 cm^{-1}, corresponding to primary NH stretching, two peaks at 2978.6 and 2900.8 cm^{-1}, which were assigned to the CH_3 asymmetric stretching vibration [42,43]. The peak near 1650 cm^{-1}, associated with the CO stretching of primary amides present in the proteins of the bacterial cell membrane. The presence of nitrile groups in the spectra is evidenced by an absorption peak at 2127 cm^{-1}, within the expected triple-bond region of 2300–2000 cm^{-1}. All of these peaks indicate significant interactions between the alginate, chitosan, and the encapsulated BB, confirming the formation of stable bonds within the microcapsule matrix. These data are consistent with findings reported by authors [44], and the amide I and II peaks are typical for proteins in microbial cells, as described in the literature [45,46].

Further analysis of the spectra shows important vibrational modes around 1460 and 1380 cm^{-1}, attributed to C-H deformation. These modes are important for confirming the structural integrity of the encapsulation matrix. The vibrational modes at 1084.5 and 1046.5 cm^{-1}, corresponding to OH deformation vibrations, are characteristic of primary (1050 cm^{-1}) and secondary alcohols (1100 cm^{-1}). These modes are closely associated with polysaccharide vibrations, aligning with previous studies on carbohydrate spectroscopy [42,47,48].

The FTIR spectrum of the encapsulated SE reveals a distinct peak at 1644.2 cm^{-1}, corresponding to the CO stretching of the carboxylic group, which results from the hydrolysis of the alginate within the encapsulation matrix, indicating interactions between SE and the alginate. Additionally, the bands observed at 1452.6 and 1402.4 cm^{-1} are attributed to CH deformation modes, further confirming the structural integrity of the encapsulated material.

The wavelength at 2120 cm^{-1} indicates the presence of isothiocyanate groups, which are known to be associated with substances found in Stevia. This observation is particularly significant as it highlights the successful encapsulation and preservation of bioactive compounds within the SE, potentially enhancing the functional properties of the encapsulated product.

The displacement of bands at 3402.7 at BB + SE microcapsules suggests the presence of amide groups, originating from the proteins in the cell membrane of BB, and effect of SE, further confirming the effective encapsulation of the active agents within the matrix [49]. The broad peaks in the range of 3000–3600 cm^{-1}, observed in these spectra, are indicative of the combined effects of O-H and N-H stretching, which overlap with polysaccharide backbone vibrations, which is consistent with the data reported by Moreno et al. [44] and other studies [50].

The appearance of three peaks at range of 2898.5–2929.8 cm^{-1} testified about interactions between BB, SE and components of encapsulating materials. The peak at 2977 cm^{-1} corresponds to the CH_3 asymmetric stretching, and is supported by a CH_2 absorption band around 2929.8 cm^{-1}, the peak at 2898.5 cm^{-1}, associated with symmetric CH_3 vibration, confirms the presence of lipid components within the membrane of the encapsulated agents, as mentioned in studies [44,51]. These lipid-associated peaks indicate the retention of critical membrane structures, which are essential for maintaining the viability and functionality of the encapsulated active agents.

The peak at 1649.8 cm^{-1}, corresponding to the CO stretching of primary amide groups, confirms the intermolecular interactions between the active agents and the encapsulation matrix composed of alginate and chitosan. This finding is in line with previous reports [42,45], where amide I and amide II bands are recognized as typical for all proteins, confirming the presence of proteinaceous components within both the BB and BB + SE.

The four small peaks at 1452.2 cm^{-1}, 1383.9 cm^{-1}, 1329.1 cm^{-1} and 1274.0 cm^{-1} suggest complex interactions between the encapsulated agents and the biopolymer matrix, which may enhance the structural integrity and stability of the microcapsules.

The peak at 1452.2 cm^{-1} corresponds to the CH_2 scissor vibration, which overlaps the CH_3 asymmetric deformation, while the peak at 1383.9 cm^{-1} corresponds to the symmetric CH_3 vibration.

The peak at 1274.0 cm^{-1} corresponds to secondary amide III vibrations, which are typically found near 1270 cm^{-1}, as reported in the literature [52]. This peak, along with others in the region of 1200–950 cm^{-1}, corresponds to the various vibrational modes of polysaccharides and polysaccharide-containing biomolecular complexes. The consistent appearance of peaks at 1280.3 cm^{-1} for BB, 1277 cm^{-1} for SE, and 1287.1 cm^{-1} for BB + SE indicates a strong correlation between these bands and the interactions within the encapsulation matrix.

The peaks at 880.0 and 670.9 correspond to the characteristic peaks of alginate and the NH deformational vibrations of chitosan, respectively. These peaks can be found in the cases of BB, SE, and BB + SE microcapsules, confirming the successful formation of an alginate-chitosan (Alg/CS) shell, resulting from interactions between oppositely charged carbonyl groups of alginate and the amino groups of chitosan. This interaction is essential for the structural integrity of the microcapsules, where the formation of polyelectrolyte complexes between alginate and chitosan was observed.

When comparing the spectra of the microcapsules containing active agents (BB, SE, and BB + SE) with those of hollow microcapsules, as described by the authors of [24], it is evident that the active agent-loaded microcapsules exhibit more intense bands. This suggests stronger intermolecular interactions and possibly a higher degree of cross-linking within the matrix. Moreover, the polysaccharide mixtures described by the authors of [50] show some spectral differences compared to the active agent-loaded microcapsules, highlighting the impact of the encapsulated agents on the overall structure and vibrational properties of the microcapsules.

The FTIR analysis of the BB + SE microcapsules reveals significant shifts in peak positions compared to the spectra of SE and BB when encapsulated separately, as well as to the spectra of hollow microcapsules.

In all three cases, distinct vibration peaks were observed, which reflect the presence of strong intermolecular interactions between the microcapsule components. These interactions contribute to the stability and functionality of the encapsulation system.

3.3. Morphological Studies

The micrographs show the morphology, size, and shape of the wet microcapsules immediately after their preparation and after drying. All wet microcapsules were almost spherical, and their surface was smooth and dense. The smooth surface was a consequence of the chitosan binding to sodium alginate chains [24,25]. Microspheres and coated microspheres are usually spherical with different surface morphologies.

Figures 2–4 display scanning electron microscope (SEM) images of the dehydrated microcapsules. The dimensions of the microcapsules observed through a scanning electron microscope (SEM) are consistent with those observed through an optical microscope (OM). The image of the microcapsules containing chitosan/alginate formulations loaded with active agents shows diameters in a range between 650 µm and 1.1 mm.

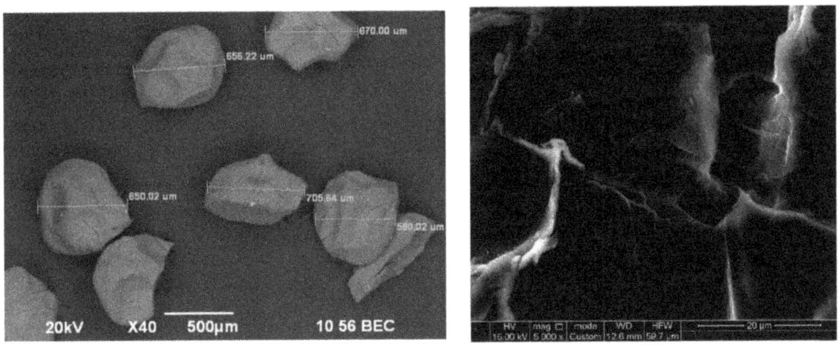

(A) (B)

Figure 2. Image of Stevia extract-loaded microcapsules: (**A**)—microcapsule diameters (at a magnification of 40); (**B**)—microcapsule porous surface (at a magnification of 5000).

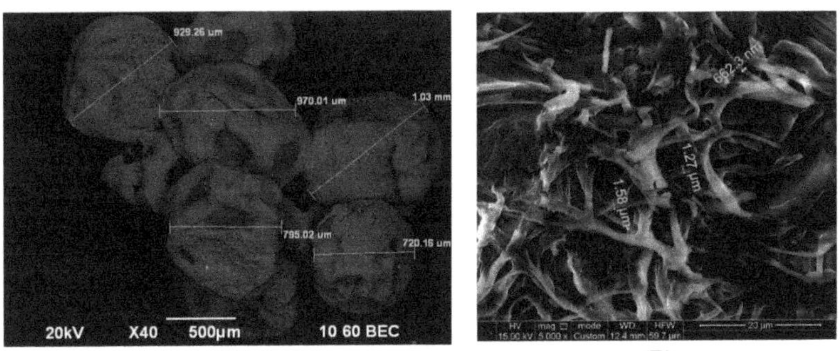

(A) (B)

Figure 3. SEM microphotograph of dry microcapsules prepared with BB: (**A**)—microcapsule diameters (at a magnification of 40); (**B**)—microcapsule porous surface (at a magnification of 5000).

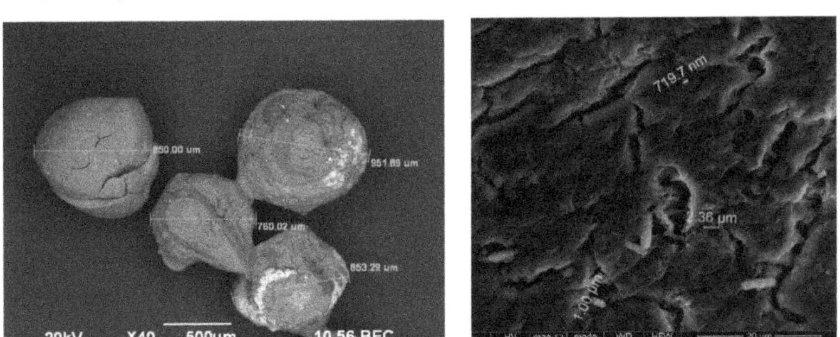

(A) (B)

Figure 4. SEM microphotograph of dry microcapsules prepared with BB + SE: (**A**)—microcapsule diameters (at a magnification of 40); (**B**)—microcapsule porous surface (at a magnification of 5000).

Figure 2 shows images of the microcapsules taken with a scanning electron microscope (SEM) after being dried for about a week at room temperature until they reached a constant mass. It is evident that they underwent a loss of their spherical morphology and significantly reduced in size compared to the wet capsules.

From Figure 2, it can be seen that the diameter of the SE-loaded microcapsules ranges from 650 to 706 μm. Figure 3 shows the BB-loaded microcapsules with different diameters.

Figure 4B shows the porous surface structure of the BB + SE-loaded dried microcapsules, with diameters of depressions on the surface varying between 719 nm and 2.36 μm. The surfaces of the microcapsules were spherical in shape and were stable when squeezed, which is consistent with data described in [24].

The color of the microcapsules (BB + SE) was yellowish, and, after drying, they became green-yellow due to the entrapped Stevia extract. In contrast, the wet microcapsules produced with BB were whitish in color, and, after drying, they developed a light yellow tone. The surfaces of the dry microcapsules were not smooth, and wrinkles could be observed. This can be explained by the fact that the produced microcapsules contained approximately 55–77% water. After the evaporation of liquid and moisture, the diameters of all prepared microcapsules decreased by half; this is associated with the relaxation–deformation process of the biopolymers, i.e., by the partial collapse of the polymer network during dehydration [53].

Produced by this method, porous dry microcapsules are suitable for living microorganisms and contribute to protection, stability and prolongation of action during use.

3.4. Release of Active Agents from Microcapsules

The release kinetics and mechanism of the active agent primarily depend on the characteristics of the core material and the active agent. For a hydrophilic biopolymer matrix, the active components are uniformly distributed/dissolved in the gel matrix, and the release is governed by diffusion, swelling, and matrix erosion as described in [24].

Figure 5 shows the release kinetics of active agents (BB, SE, and BB + SE) from microcapsules as a function of time exhibited a continuous and gradual release pattern. Within the initial period of two hours, approximately 38% of BB and SE were released, which increased to approximately 55–58% after four hours. This can be assumed a steady diffusion of the active agents through the gel matrix. For BB + SE, as a combination of active agents, the release showed faster kinetics. Within two hours, around 60% of the active agents were released, and this value further increased to 70% after four hours. This accelerated release may be explained by a synergistic effect between BB and SE, which could enhance the permeability of the encapsulation matrix or alter the internal microenvironment, facilitating a quicker diffusion of the active agents. After 24 h, BB + SE combination were released almost completely, indicating that the microcapsules were effective in providing a required release, ensuring that a significant amount of the active agents are available at the desired period.

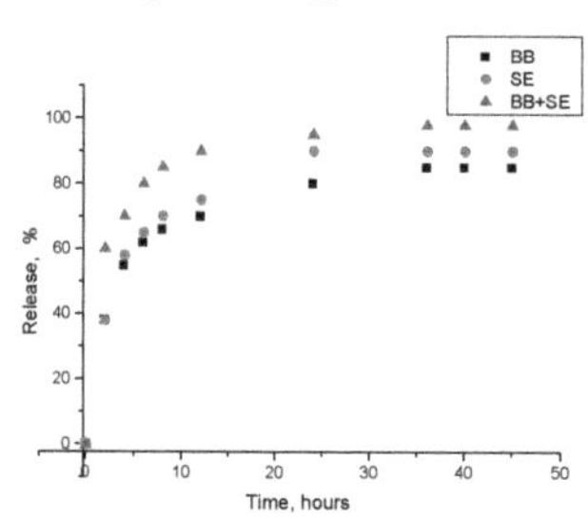

Figure 5. Release kinetics of active agents over time: BB, SE, and BB + SE.

From the literature, it is known that food generally stays in the stomach for between 40 min and 2 h. It then spends around five hours in the small intestine, before passing through the colon, which can take between 10 and 59 h [54]. Thus, release over 24 h is enough for probiotics' survival.

Thus, research findings of release kinetics show that microcapsules produced by loading of a combination of BB + SE can have controlled release.

3.5. Behavior of Active Agents Loaded Microcapsules in an Imitating Gastrointestinal Tract

The next release percentage $R_{(t)}$ using piecewise linear function have been calculated: $R_{BB(t)} = R_{SE(t)} = R_{(BB + SE)(t)}$: for $2 \leq t \leq 4$ and for $4 \leq t \leq 24$.

Information about swelling behavior of the active agent loaded microcapsules have been modeled using an exponential function in dependence of time and pH of GIT as presented in Table 3.

Table 3. Parameters for swelling behavior of microcapsules in different sections of the GIT.

Section of GIT	pH	k	α
Stomach	1–3	0.5 ± 0.1	0.15 ± 0.0361
Small intestine	6–7	0.3 ± 0.1	0.1 ± 0.0361
Colon intestine	6–7.5	0.4 ± 0.1	0.08 ± 0.0361

Modeling of swelling behavior at different GIT sections has shown that in the stomach section the $\alpha = 0.15$ testified about rapid swelling because of high acidity. In the small intestine, this value is equal to 0.10, which confirms moderate swelling, and $\alpha = 0.08$ testified about slow release.

Design of release kinetics and swelling behavior show that the developed microcapsules are suitable for the controlled release of BB, SE and BB + SE in a simulated GI tract. Initially, there is a rapid release in the stomach, which is associated with high acidity. This is followed by a slow release in the small and colon intestines, which provides optimal conditions for probiotic survival and activity. Thus, the design results show that the developed microcapsules are effective in application.

4. Conclusions

The study results successfully demonstrated the potential of encapsulating *Bifidumbacteria* (BB) and stevia extract (SE) into alginate-chitosan microcapsules for use in fermented beverage fortification. FTIR analysis confirmed significant interactions between the encapsulated agents and the biopolymer matrix, indicating the formation of a stable and efficient encapsulation system. The high encapsulation efficiencies observed (83.0% for BB, 89.2% for SE, and 91.3% for BB + SE) reflect the reliability of the system for protection and enhancing the survivability of active agents in the aggressive environment. Particularly for the co-encapsulation of BB and SE, the synergistic effect likely enhanced the stability and retention of the active agents within microcapsules.

The swelling degree of the capsules loaded with BB was 79%, that of the capsules loaded with SE was 73%, and that of the capsules loaded with BB + SE was 53%. This lower swelling and moisture content indicate a more compact and stable structure. This compact structure was further supported by the observed wrinkled porous surface morphology, resulting from the partial collapse of the polymer network during the drying process.

The morphology of the dry capsules showed that there were cracks on the surface, with diameters between 719.7 nm and 2.36 μm. In addition, there were recesses, with widths between 81 and 127 μm and lengths from 162 to 384 μm.

The release kinetics of the microcapsules demonstrated a controlled and sustained release over 24 h, aligning with the transit time of food through the gastrointestinal tract. The faster initial release observed in the BB + SE microcapsules suggests potential synergistic interactions that could be advantageous in applications requiring a release at desired time.

Modeling of release kinetics and swelling behavior of microcapsules in the imitating GIT show that the developed microcapsules are suitable for the protection of active agents at application in the enrichment of fermented beverage.

The encapsulation of probiotic bacteria and plant extracts lays the groundwork for innovative applications in the fortification of fermented beverages. These beverages will be developed using acid whey, a byproduct of the dairy industry that is notably rich in essential micro- and macro-elements. By repurposing acid whey, this research not only contributes to waste reduction but also enhances the nutritional value of the resulting fermented products.

The novelty of this investigated encapsulation system is the increased stability, bioavailability, and functionality of the active agents, which provides a foundation for the development of functional fermented beverages enriched with probiotics and plant extracts, offering health benefits for consumers.

Author Contributions: Conceptualization, B.M. and E.T.; methodology, E.T. and B.M.; software, A.I. and A.S.; validation, G.M., B.M. and S.A.; formal analysis, B.M., E.T.; investigation, E.T., D.O., G.M. and B.M.; resources, B.M. and G.M.; data curation, B.M., E.T., G.M.; writing—original draft preparation, E.T. and B.M.; writing—review and editing, B.M., R.M., G.M. and A.I.; visualization, E.T., B.M. and R.M.; supervision, B.M.; project administration, G.M. and B.M.; funding acquisition, G.M. All authors have read and agreed to the published version of the manuscript.

Funding: This research is funded by the Science Committee of the Ministry of Science and Higher Education of the Republic of Kazakhstan (Grant No. AP19679879).

Data Availability Statement: The original contributions presented in the study are included in the article, further inquiries can be directed to the corresponding author/s.

Conflicts of Interest: The authors declare no conflicts of interest.

References

1. Solieri, L.; Valentini, M.; Cattivelli, A.; Sola, L.; Helal, A.; Martini, S.; Tagliazucchi, D. Fermentation of whey protein concentrate by Streptococcus thermophilus strains releases peptides with biological activities. *Process. Biochem.* **2022**, *121*, 590–600. [CrossRef]
2. Jitpakdee, J.; Kantachote, D.; Kanzaki, H.; Nitoda, T. Potential of lactic acid bacteria to produce functional fermented whey beverage with putative health promoting attributes. *LWT* **2022**, *160*, 113269. [CrossRef]
3. Skryplonek, K.; Dmytrów, I.; Mituniewicz-Małek, A. Probiotic fermented beverages based on acid whey. *J. Dairy Sci.* **2019**, *102*, 7773–7780. [CrossRef] [PubMed]
4. Shi, X.; Wu, D.; Xu, Y.; Yu, X. Engineering the optimum pH of β-galactosidase from Aspergillus oryzae for efficient hydrolysis of lactose. *J. Dairy Sci.* **2022**, *105*, 4772–4782. [CrossRef]
5. Wherry, B.; Barbano, D.M.; Drake, M.A. Use of acid whey protein concentrate as an ingredient in nonfat cup set-style yogurt. *J. Dairy Sci.* **2019**, *102*, 8768–8784. [CrossRef] [PubMed]
6. Molero, M.S.; Briñez, W.J. Probiotics Consumption Increment through the Use of Whey-Based Fermented Beverages. In *Probiotics—Current Knowledge and Future Prospects*; InTech: Houston, TX, USA, 2018. [CrossRef]
7. Food and Agriculture Organization of the United Nations; World Health Organization. *Guidelines for the Evaluation of Probiotics in Food*; Food and Agriculture Organization of the United Nations: Rome, Italy; World Health Organization: Geneva, Switzerland, 2002.
8. Saad, S.M.I. Probiotics and prebiotics: The state of the art. *Braz. J. Pharm. Sci.* **2006**, *42*, 1–16.
9. Mgomi, F.C.; Yuan, L.; Farooq, R.; Lu, C.-L.; Yang, Z.-Q. Survivability and characterization of the biofilm-like probiotic Pediococcus pentosaceus encapsulated in calcium alginate gel beads. *Food Hydrocoll.* **2024**, *156*, 110253. [CrossRef]
10. Shori, A.B. Microencapsulation improved probiotics survival during gastric transit. *J. Biosci.* **2017**, *24*, 1–5. [CrossRef]
11. da Silva, M.N.; Tagliapietra, B.L.; Flores, V.D.A.; Richards, N.S.P.d.S. In vitro test to evaluate survival in the gastrointestinal tract of commercial probiotics. *Curr. Res. Food Sci.* **2021**, *4*, 320–325. [CrossRef]
12. Champagne, C.P.; Gomes da Cruz, A.; Daga, M. Strategies to improve the functionality of probiotics in supplements and foods. *Curr. Opin. Food Sci.* **2018**, *22*, 160–166. [CrossRef]
13. Krunić, T.Ž.; Rakin, M.B. Enriching alginate matrix used for probiotic encapsulation with whey protein concentrate or its trypsin-derived hydrolysate: Impact on antioxidant capacity and stability of fermented whey-based beverages. *Food Chem.* **2022**, *370*, 130931. [CrossRef] [PubMed]
14. Zanjani, M.; Tarzia, B.; Sharifana, A.; Mohammadi, N. Microencapsulation of Probiotics by Calcium Alginate-gelatinized Starch with Chitosan Coating and Evaluation of Survival in Simulated Human Gastro-intestinal Condition. *Iran J. Pharm. Res. (IJPR)* **2014**, *13*, 843–852.
15. Yao, M.; Xie, J.; Du, H.; McClements, D.J.; Xiao, H.; Li, L. Progress in microencapsulation of probiotics: A review. *Compr. Rev. Food Sci. Food Saf.* **2020**, *19*, 857–874. [CrossRef] [PubMed]
16. Alemzadeh, E.; Oryan, A. Application of encapsulated probiotics in health care. *J. Exp. Path.* **2020**, *1*, 16–21.
17. Ding, W.K.; Shah, N.P. Effect of Various Encapsulating Materials on the Stability of Probiotic Bacteria. *J. Food Sci.* **2009**, *74*, M100–M107. [CrossRef]

18. Gaonkar, A.; Vasisht, N.; Khare, A.; Sobel, R. *Microencapsulation in the Food Industry*, 1st ed.; A Practical Implementation Guide; Academic Press: Cambridge, MA, USA, 2014.
19. Ozdal, T.; Yolci-Omeroglu, P.; Tamer, E.C. 6-Role of Encapsulation in Functional Beverages. In *Biotechnological Progress and Beverage Consumption*; Grumezescu, A.M., Holban, A.M., Eds.; Academic Press: Cambridge, MA, USA, 2020; pp. 195–232. [CrossRef]
20. Nezamdoost-Sani, N.; Khaledabad, M.A.; Amiri, S.; Mousavi Khaneghah, A. Alginate and derivatives hydrogels in encapsulation of probiotic bacteria: An updated review. *Food Biosci.* **2023**, *52*, 102433. [CrossRef]
21. Amiri, S.; Abotalebi Kohneshahri, S.R.; Nabizadeh, F. The effect of unit operation and adjunct probiotic culture on physicochemical, biochemical, and textural properties of Dutch Edam cheese. *LWT* **2022**, *155*, 112859. [CrossRef]
22. Thambiliyagodage, C.; Jayanetti, M.; Mendis, A.; Ekanayake, G.; Liyanaarachchi, H.; Vigneswaran, S. Recent Advances in Chitosan-Based Applications—A Review. *Materials* **2023**, *16*, 2073. [CrossRef]
23. Xie, A.; Zhao, S.; Liu, Z.; Yue, X.; Shao, J.; Li, M.; Li, Z. Polysaccharides, proteins, and their complex as microencapsulation carriers for delivery of probiotics: A review on carrier types and encapsulation techniques. *Int. J. Biol. Macromol.* **2023**, *242*, 124784. [CrossRef]
24. Vlahoviček-Kahlina, K.; Jurić, S.; Marijan, M.; Mutaliyeva, B.; Khalus, S.V.; Prosyanik, A.V.; Vinceković, M. Synthesis, Characterization, and Encapsulation of Novel Plant Growth Regulators (PGRs) in Biopolymer Matrices. *Int. J. Mol. Sci.* **2021**, *22*, 1847. [CrossRef]
25. Kudasova, D.; Mutaliyeva, B.; Vlahoviček-Kahlina, K.; Jurić, S.; Marijan, M.; Khalus, S.V.; Prosyanik, A.V.; Šegota, S.; Španić, N.; Vinceković, M. Encapsulation of Synthesized Plant Growth Regulator Based on Copper(II) Complex in Chitosan/Alginate Microcapsules. *Int. J. Mol. Sci.* **2021**, *22*, 2663. [CrossRef]
26. Ta, L.P.; Bujna, E.; Antal, O.; Ladányi, M.; Juhász, R.; Szécsi, A.; Kun, S.; Sudheer, S.; Gupta, V.K.; Nguyen, Q.D. Effects of various polysaccharides (alginate, carrageenan, gums, chitosan) and their combination with prebiotic saccharides (resistant starch, lactosucrose, lactulose) on the encapsulation of probiotic bacteria Lactobacillus casei 01 strain. *Int. J. Biol. Macromol.* **2021**, *183*, 1136–1144. [CrossRef] [PubMed]
27. Peredo, A.; Beristain, C.; Pascual, L.; Azuara, E.; Jimenez, M. The effect of prebiotics on the viability of encapsulated probiotic bacteria. *LWT* **2016**, *73*, 191–196. [CrossRef]
28. Ismail, S.A.; Hassan, A.A.; Nour, S.A.; El-Sayed, H.S. The production of stirred yogurt fortified with prebiotic xylooligosaccharide, probiotic and synbiotic microcapsules. *Biocatal. Agric. Biotechnol.* **2023**, *50*, 102729. [CrossRef]
29. Zhou, Z.; Sarwar, A.; Xue, R.; Hu, G.; Wu, J.; Aziz, T.; Alasmari, A.F.; Yang, Z.; Yang, Z. Metabolomics analysis of potential functional metabolites in synbiotic ice cream made with probiotic Saccharomyces cerevisiae var. boulardii CNCM I-745 and prebiotic inulin. *Food Chem.* **2024**, *454*, 139839. [CrossRef] [PubMed]
30. Prakash, K.S.; Bashir, K.; Mishra, V. Development of symbiotic litchi juice drink and its physiochemical, viability and sensory analysis. *J. Food Process. Technol.* **2017**, *8*, 1–6. [CrossRef]
31. Kurek, J.M.; Krejpcio, Z. The functional and health-promoting properties of Stevia rebaudiana Bertoni and its glycosides with special focus on the antidiabetic potential–A review. *J. Funct. Foods* **2019**, *61*, 103465. [CrossRef]
32. Jungersen, M.; Wind, A.; Johansen, E.; Christensen, J.E.; Stuer-Lauridsen, B.; Eskesen, D. The Science behind the Pro-biotic Strain Bifidobacterium Animalis Subsp. Lactis BB-12®. *Microorganisms* **2014**, *2*, 92–110. [CrossRef]
33. Matera, M. *Bifidobacteria, Lactobacilli... When, How and Why to Use Them*; Global Pediatrics: New York, NY, USA, 2024; Volume 8, p. 100139. [CrossRef]
34. Pyle, S. 2.23-Human Gut Microbiota and the Influence of Probiotics, Prebiotics, and Micronutrients. *Compr. Gut Microbiota* **2022**, *2*, 271–288. [CrossRef]
35. Voblikova, T. Viability of the culture of Bifidobacterium bifidum immobilized by microencapsulation in dairy drink and the simulated gastrointestinal liquids. *Vestnik MGTU* **2019**, *22*, 305–313. [CrossRef]
36. Hamman, J. Chitosan Based Polyelectrolyte Complexes as Potential Carrier Materials in Drug Delivery Systems. *Mar Drugs* **2010**, *8*, 1305–1322. [CrossRef] [PubMed]
37. Higuchi, T. Mechanism of sustained-action medication. Theoretical analysis of rate of release of solid drugs dispersed in solid matrices. *J. Pharm. Sci.* **1963**, *52*, 1145–1149. [CrossRef]
38. Fu, J.; Cai, Y.; Le, Q.; Su, Z. Controlled release behavior of a pH-sensitive hydrogel microsphere based on modified alginate. *Int. J. Biol. Macromol.* **2018**, *112*, 218–225. [CrossRef]
39. Aguilera, D.A.; Di Sante, L.S.; Pettignano, A.; Riccioli, R.; Roeske, J.; Albergati, L.; Corti, V.; Fochi, M.; Bernardi, L.; Quignard, F.; et al. Adsorption of a Chiral Amine on Alginate Gel Beads and Evaluation of its Efficiency as Heterogeneous Enantioselective Catalyst. *Eur. J. Org. Chem.* **2019**, *2019*, 3842–3849. [CrossRef]
40. Xu, C.; Ban, Q.; Wang, W.; Hou, J.; Jiang, Z. Novel nano-encapsulated probiotic agents: Encapsulate materials, delivery, and encapsulation systems. *J. Control. Release* **2022**, *349*, 184–205. [CrossRef]
41. Burgain, J.; Corgneau, M.; Scher, J.; Gaiani, C. *Chapter 20—Encapsulation of Probiotics in Milk Protein Microcapsules*; Academic Press: Cambridge, MA, USA, 2015. [CrossRef]
42. Kamnev, A.A.; Dyatlova, Y.A.; Kenzhegulov, O.A.; Vladimirova, A.A.; Mamchenkova, P.V.; Tugarova, A.V. Fourier Transform Infrared (FTIR) Spectroscopic Analyses of Microbiological Samples and Biogenic Selenium Nanoparticles of Microbial Origin: Sample Preparation Effects. *Molecules* **2021**, *26*, 1146. [CrossRef]

43. Kamnev, A.A.; Dyatlova, Y.A.; Kenzhegulov, O.A.; Fedonenko, Y.P.; Evstigneeva, S.S.; Tugarova, A.V. Fourier Transform Infrared (FTIR) Spectroscopic Study of Biofilms Formed by the Rhizobacterium Azospirillum baldaniorum Sp245: Aspects of Methodology and Matrix Composition. *Molecules* **2023**, *28*, 1949. [CrossRef]
44. Moreno, J.S.; Dima, P.; Chronakis, I.S.; Mendes, A.C. Electrosprayed Ethyl Cellulose Core-Shell Microcapsules for the Encapsulation of Probiotics. *Pharmaceutics* **2022**, *14*, 7. [CrossRef]
45. Naumann, D. Infrared Spectroscopy in Microbiology. In *Encyclopedia of Analytical Chemistry*; Wiley: Hoboken, NJ, USA, 2006. [CrossRef]
46. Barth, A. Infrared spectroscopy of proteins. *Biochim. Biophys. Acta Bioenerg.* **2007**, *1767*, 1073–1101. [CrossRef]
47. Naumann, D.; Helm, D.; Labischinski, H. Microbiological characterizations by FT-IR spectroscopy. *Nature* **1991**, *351*, 81–82. [CrossRef]
48. Wiercigroch, E.; Szafraniec, E.; Czamara, K.; Pacia, M.Z.; Majzner, K.; Kochan, K.; Kaczor, A.; Baranska, M.; Malek, K. Raman and infrared spectroscopy of carbohydrates: A review. *Spectrochim. Acta Part A Mol. Biomol. Spectrosc.* **2017**, *185*, 317–335. [CrossRef]
49. Gordon, S.H.; Harry-O'kuru, R.E.; Mohamed, A.A. Elimination of interference from water in KBr disk FT-IR spectra of solid biomaterials by chemometrics solved with kinetic modeling. *Talanta* **2017**, *174*, 587–598. [CrossRef] [PubMed]
50. Baysal, K.; Aroguz, A.; Adiguzel, Z.; Baysal, B. Chitosan/Alginate Crosslinked Hydrogels: Preparation, Characterization and Application for Cell Growth Purposes. *Int. J. Biol. Macromol.* **2013**, *59*, 342–348. [CrossRef] [PubMed]
51. Vodnar, D.; Adriana, P.; Francisc, D.; Socaciu, C. HPLC Characterization of Lactic Acid Formation and FTIR Fingerprint of Probiotic Bacteria during Fermentation Processes. *Not. Bot. Horti Agrobot. Cluj Napoca* **2010**, *38*, 109–113.
52. Yin, M.; Chen, L.; Chen, M.; Yuan, Y.; Liu, F.; Zhong, F. Encapsulation of Lactobacillus rhamnosus GG in double emulsions: Role of prebiotics in improving probiotics survival during spray drying and storage. *Food Hydrocoll.* **2024**, *151*, 109792. [CrossRef]
53. Jurić, S.; Šegota, S.; Vinceković, M. Influence of surface morphology and structure of alginate microparticles on the bioactive agents release behavior. *Carbohydr. Polym.* **2019**, *218*, 234–242. [CrossRef]
54. Mackie, A.; Mulet-Cabero, A.-I.; Torcello-Gómez, A. Simulating human digestion: Developing our knowledge to create healthier and more sustainable foods. *Food Funct.* **2020**, *11*, 9397–9431. [CrossRef]

colloids and interfaces

MDPI

Article

Impact of Carrageenan-Based Encapsulation on the Physicochemical, Structural, and Antioxidant Properties of Freshwater Snail (*Bellamya bengalensis*) Protein Hydrolysates

Anand Vaishnav [1], Naresh Kumar Mehta [1,*], Mocherla Bhargavi Priyadarshini [1], Soibam Khogen Singh [2], Pratap Chandra Acharya [3], Satyajeet Biswal [3], Harjeet Nath [4], Syed Arshad Hussain [5], Prasenjit Pal [6], Jham Lal [7], Nongthongbam Sureshchandra Singh [1] and Bikash Kumar Pati [1]

1 Department of Fish Processing Technology and Engineering, College of Fisheries, Central Agricultural University (Imphal), Lembucherra 799210, Tripura, India; anandcof9150@gmail.com (A.V.); mocherla.bhargavi@gmail.com (M.B.P.); sureshnong16@gmail.com (N.S.S.); bikashprocessing@gmail.com (B.K.P.)

2 Krishi Vigyan Kendra, Indian Council of Agricultural Research, Regional Centre, NEH Region, Ukhrul 795142, Manipur, India; gengang@gmail.com

3 Drug Metabolomics Laboratory, Department of Pharmacy, Tripura University (A Central University), Suryamaninagar 799022, Tripura, India; pratapacharya@tripurauniv.ac.in (P.C.A.); mail.sbiswal@gmail.com (S.B.)

4 Department of Chemical and Polymer Engineering, Tripura University (A Central University), Suryamaninagar 799022, Tripura, India; harjeetnath@tripurauniv.ac.in

5 Department of Physics, Tripura University (A Central University), Suryamaninagar 799022, Tripura, India; sahussain@tripurauniv.ac.in

6 Department of Fisheries Extension, Economics and Statistics, College of Fisheries, Central Agricultural University (Imphal), Lembucherra 799210, Tripura, India; prasenjit3agstat@gmail.com

7 Department of Aquaculture, College of Fisheries, Central Agricultural University (Imphal), Lembucherra 799210, Tripura, India; jhamlalj@gmail.com

* Correspondence: nareshfishco@gmail.com; Tel.: +91-9967088756

Abstract: This study investigated the encapsulation of snail protein hydrolysates (SPHs) using carrageenan as a microencapsulating agent at concentrations of 1%, 2%, and 3%. SPHs were prepared from the soft tissue of freshwater snails (*Bellamya bengalensis*) through enzymatic hydrolysis using bromelain, resulting in a degree of hydrolysis of 48.05%. The encapsulation process was carried out using the spray-drying technique. Encapsulation with 3% carrageenan enhanced the yield, encapsulation efficiency (up to 84.96%), colloidal stability (up to −33.8 mV), and thermal stability (up to 75 °C). The particle size increased as the carrageenan concentration increased, reaching 206.9 nm at 3%, and the uniform polydispersity index (0.26) indicated stable encapsulation. While encapsulation reduces solubility and antioxidant activity (DPPH, FRAP, ABTS, and HRSA), it effectively protects SPH from environmental factors such as hygroscopicity and storage stability, thus maintaining high scavenging activity. Fourier transform infrared spectroscopy confirmed that carrageenan and SPH strongly interact. Scanning electron microscopy revealed that the particles had better shapes and smooth, cohesive surfaces. This study demonstrates the effectiveness of carrageenan as an encapsulating agent for SPH, enhancing its stability and bioactivity for potential applications in the food and nutraceutical industries as a bioactive additive and offering an alternative to conventional coating materials.

Keywords: carrageenan; colloidal stability; particle size; thermal stability; freshwater snail; microencapsulation

1. Introduction

The global demand for functional food ingredients with bioactive properties has spurred interest in protein hydrolysates derived from diverse natural sources because of their multifaceted health benefits and functional properties [1,2]. This quest has led to the possibility of discovering diverse biological resources, including aquatic organisms. Researchers have targeted various aquatic sources, such as fresh and marine water fishes, shellfishes, seaweeds, and processing waste from aquatic foods, to produce bioactive peptides [3–5]. However, freshwater snails are often less explored despite their high protein content, essential amino acid content, and abundance in many regions of the world [6]. In addition to their potential nutritional and health benefits, snails are a cheap, sustainable, and rich source of bioactive compounds that could have positive effects on human health [7, 8]. The integration of freshwater snails into food products may expand dietary choices and facilitate advancements in functional foods and nutraceuticals. They are excellent choices for making protein hydrolysates because of their high protein (13–17%) content [9,10] and essential amino acids such as lysine, leucine, histidine, and valine [11,12]. These snail-derived compounds potentially offer various health benefits, thus opening new possibilities for the development of health-promoting food products [13,14]. Protein hydrolysates derived via enzymatic hydrolysis are recognized for their bioactive peptides [15,16]. The facile digestion of peptides renders them optimal components for food items [17]. There is an increasing trend in the production of functional foods and nutraceuticals fortified with peptides.

Despite these advantages, their compatibility may be influenced by challenges such as susceptibility to degradation, oxidation, diminished bioactivity during storage and processing, undesirable bitter flavor, hygroscopicity, hydrophobicity, chemical instability, and decreased interactions with the food matrix [18,19]. Furthermore, obstacles such as restricted bioavailability and biostability hinder their successful integration into food. A viable solution to address these issues is the microencapsulation of protein hydrolysates via methods such as spray drying [20,21]. Microencapsulation is a method for enclosing bioactive substances in a protective matrix. Recently, many methods have been developed to improve the stability and functionality of sensitive compounds, making controlled release easier and making them more useful in a wider range of situations [22,23]. Spray drying is a prevalent method for microencapsulation in the food and nutraceutical sectors because of its efficiency, scalability, and cost-effectiveness [24]. This process involves atomizing a liquid mixture including the core material (e.g., bioactive compounds or protein hydrolysates) and the encapsulating material into fine droplets and subsequently subjecting it to fast drying with hot air to yield stable, dry particles [25]. Spray drying is beneficial for sensitive bioactive compounds since it reduces thermal degradation, enables precise control over particle size, extends shelf-life, and facilitates the incorporation of these powders into food products [26].

Among various encapsulation matrices, carrageenan, a natural, water-soluble, negatively charged galactose-based polysaccharide derived from red algae (Rhodophyceae), has shown excellent potential because its biocompatibility, gel-forming properties, and ability to form stable microcapsules through thermal or ion-induced gelation (hydrogel formation) make it suitable for this application [27]. In contrast to maltodextrin, gum arabic, or alginate, carrageenan provides a more robust gel network and enhanced moisture barrier characteristics, particularly beneficial for encapsulating animal-derived protein hydrolysates susceptible to oxidation and hygroscopicity. Its anionic characteristics facilitate electrostatic interactions with peptides, enhancing structural stability and prolonged release [27]. Furthermore, carrageenan conforms to the clean-label approach owing to its natural provenance and is more cost-effective and thermally stable than gum arabic, maltodextrin and alginate, rendering it a strategic option for food-grade encapsulation of bioactive peptides [28].

It has preventive benefits, enhances bioavailability, and corresponds with the increasing desire for clean-label products that are considered healthful and environmentally sustainable [28,29]. There are still relatively few investigations into carrageenan-based encapsulation techniques for animal-derived protein hydrolysates, as the majority of research has focused on plant-derived proteins. This gap underscores the need for a more comprehensive understanding of the effectiveness of carrageenan in encapsulating animal-derived bioactive peptides, especially from underused freshwater snail species such as *Bellamya bengalensis*. Carrageenan can be used for encapsulation through spray drying of numerous bioactive substances, such as antioxidants, vitamins, and essential oils, increasing their stability and function in food systems [30,31].

This process produces ultrafine particles that act as protective barriers, effectively protecting sensitive ingredients from exposure to environmental factors such as light, heat, and oxygen [32]. Using carrageenan as an encapsulating agent can also make some bioactive compounds more palatable in food products. This is because it masks the unpleasant taste or smell that some bioactive compounds have, making them more palatable [31]. Although encapsulated protein hydrolysates have potential for use in functional foods and nutraceuticals, studies on encapsulating protein hydrolysates from freshwater snails are scarce. Research has not yet thoroughly examined the effects of carrageenan encapsulation on the physicochemical and antioxidant characteristics of SPH, especially the improvement in shelf stability and the retention of bioactive properties post microencapsulation.

Therefore, the objective of this study is to evaluate the impact of carrageenan-based spray-drying encapsulation at varying concentrations (1%, 2%, and 3%) on the physicochemical, structural, and antioxidant properties of protein hydrolysates derived from freshwater snail (*Bellamya bengalensis*).

2. Materials and Methods

2.1. Chemicals

The *B. bengalensis* snails used in the experiment were purchased from a nearby market in Agartala, Tripura, India, and then immersed in filtered water for a period of 48 h to remove any contaminants. Before being frozen at -18 ± 2 °C for additional analysis, the meat was manually separated, rinsed, and analyzed. The microencapsulation agent's carrageenan and pineapple stem tissue-derived bromelain (0.5 DMC U/mg) were purchased from HiMedia Ltd. (Thane (West), India) and stored in a refrigerator.

2.2. Preparation of Protein Hydrolysate

Snail protein hydrolysates (SPHs) were prepared according to Vaishnav et al. [33], with specific modifications. Figure 1 shows the detailed procedure of SPH preparation. Snail soft tissue was thawed at 27 ± 3 °C, sliced, and ground into a paste. Flash paste was mixed with double distilled water at a 1:1 ratio and homogenized at low speed for 2 min. The mixture was heated to 85 °C for 20 min to inactivate endogenous enzymes and then cooled, and the pH was adjusted to 7 with 0.1 N NaOH/HCl solutions. The material was subsequently agitated via a magnetic stirrer at 50 °C for hydrolysis. Bromelain was added at a 1:100 ratio (*w*/*w*) to initiate hydrolysis, and the pH was meticulously controlled. Following hydrolysis, the mixture was heated to 90 °C for 20 min to deactivate bromelain. After centrifugation at 8000 rpm for 10 min, the supernatant was collected and stored at -20 °C for future use.

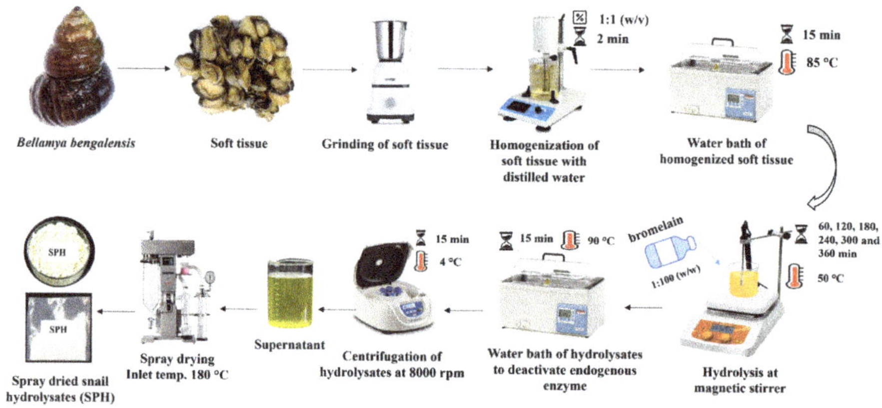

Figure 1. Preparation of protein hydrolysates from freshwater snails (*B. bengalensis*).

2.3. Degree of Hydrolysis (DH)

The method outlined by Hoyle and Merritt [34] was employed to assess the degree of hydrolysis (DH). The DH was assessed at intervals of 60, 120, 180, 240, 300, and 360 min. At each time interval, a sample was collected and combined with 20% (1:1 ratio) trichloroacetic acid (TCA) to inactivate the enzymes. The mixture was then filtered, and the soluble protein content in the filtrate was analyzed. The DH was calculated as a ratio of the total protein content in the substrate, which was determined via the Kjeldahl method [35], to the amount of soluble proteins in the filtrate, which was measured via the Lowry method [36].

2.4. Encapsulation of Hydrolysates

Snail protein hydrolysates (SPH) were prepared through enzymatic hydrolysis (as described in the preceding Section 2.2) and subsequently subjected to microencapsulation using carrageenan as the wall material. Carrageenan was mixed with snail protein hydrolysate at concentrations of 1% (CN 1%), 2% (CN 2%), and 3% (CN 3%) (w/v) (as shown in Figure 2) and stirred on a magnetic stirrer (Neuation iStir HP 320, Gandhinagar, India) at 150 rpm for 10–15 min to ensure proper mixing. The mixtures were then spray-dried (SprayMate LSD-48, Jay Instruments & System Pvt. Ltd., Navi Mumbai, India) by pumping them into the spray dryer at a rate of 500 mL/h with a peristaltic pump. An atomizer operating at 1400 rpm was utilized for atomization in a downward flow arrangement, with an inlet temperature of 180 °C and an outflow temperature of 60 ± 3 °C. The resulting powder was separated via a cyclone separator of spray drier and collected in a container. The powdered samples were packed and stored in vacuum-sealed plastic bags composed of multilayer polyethylene (PE) with a total thickness of 100 µm. This type of packaging offers low oxygen and moisture permeability, making it suitable for long-term storage of sensitive bioactive compounds. The sealed samples were stored at −20 °C until further investigation.

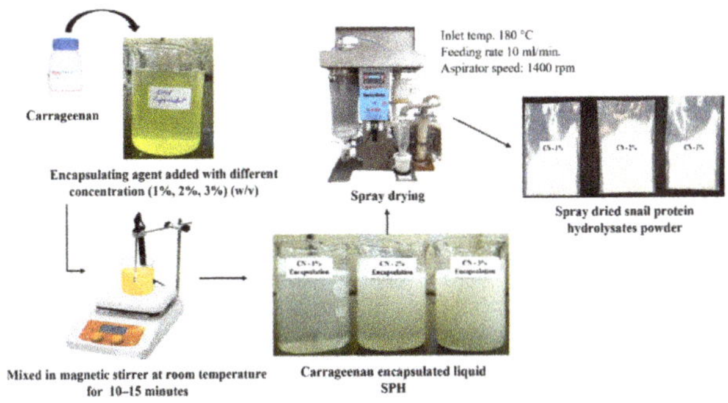

Figure 2. Encapsulation of snail protein hydrolysates with 1%, 2%, and 3% carrageenan.

2.5. *Yield*

The yield of the SPH and encapsulated SPH were determined by calculating the ratio of the mass of spray-dried hydrolysate powder to the initial mass of the raw material used for hydrolysis. The yield was expressed as a percentage, following the method described by Elavarasan et al. [37].

2.6. *Encapsulation Efficiency (EE)*

The encapsulation efficiency of the encapsulated protein hydrolysates was determined via the method described by Mendanha et al. [38]. The encapsulated hydrolysate powder was dissolved in water (5 mg/mL) and centrifuged to separate the nonencapsulated hydrolysates from the encapsulated particles. The amount of nonencapsulated hydrolysates in the supernatant was then quantified via the biuret method. The protein content of the encapsulated hydrolysates was also calculated. The encapsulation efficiency was calculated via the following formula:

$$\text{EE}\ (\%) = \frac{\text{Total protein content} - \text{Encapsulated protein content}}{\text{Total protein content}} \times 100 \quad (1)$$

2.7. *Hygroscopicity*

To determine moisture sorption, the method described by Unnikrishnan et al. [39] was used, and preweighed dry samples were placed in sealed desiccators containing a saturated sodium chloride solution, which maintained a relative humidity of 75.3% at 25 °C. After one week, the samples were removed and reweighed. The hygroscopicity was calculated as the percentage increase in weight due to moisture absorption relative to the initial dry weight of the samples.

2.8. *Color Analysis*

The color of the snail protein hydrolysates (SPHs) and encapsulated hydrolysates was measured via a colorimeter (ColorFlex, EZ, Hunter Associates Laboratory, Reston, VA, USA). The instrument measured standard color parameters, including lightness (L^*), redness (a^*), and yellowness (b^*). The whiteness and chroma were calculated as described by Zaitounet al. [40].

2.9. *Amino Acid Profiling*

The amino acid profile of SPH was analyzed via high-resolution mass spectrometry (HRMS) with a Q-TOF mass spectrometer (6200 series TOF/6500 series Q-TOF, Agilent Technologies, Gurgaon, India) using the method described by Yin et al. [41]. A 2-µL aliquot of the sample was analyzed for amino acids, utilizing the amino acid MS method with

MassHunter software (Version 12.1) for data acquisition and analysis. During profiling, individual amino acids were identified by their retention times, molecular formulas, and mass—charge ratios (m/z) against a reference database. Mass differences were also calculated to confirm accuracy, and compound identification was assigned a score on the basis of the reliability of the match.

2.10. Particle Size Distribution

The particle size distributions of the spray-dried SPH and encapsulated SPH samples were assessed via dynamic light scattering (Anton Paar, Graz, Austria). The samples were prepared under standardized conditions and evaluated to determine the size distribution and polydispersity index. The measurements were conducted at 25 °C with a scattering angle of 90°, and the particle sizes were measured after 120 s of autocorrelation.

2.11. Zeta Potential Measurements

Zeta potential measurements were conducted with Anton Paar (Austria) equipment to assess the surface charge of the hydrolysate particles. The electrophoretic mobility of the particles was assessed using laser doppler velocimetry technique. Each sample was measured thrice at 20 °C to ensure precision in the acquired zeta potential values.

2.12. Solubility

This method of Klompong et al. [42] was used to measure protein solubility. Two hundred milligrams of spray-dried SPH and encapsulated SPH were dissolved in 20 mL of deionized water. HCl (1 N) and NaOH (1 N) were used to adjust the solution pH to 2, 4, 6, 8, or 10. After 30 min of stirring at room temperature, the mixture was centrifuged at 7500× g for 15 min. The protein content of the supernatant was measured using the Biuret method [43]. The material was diluted in 0.5 N NaOH for protein analysis. Protein solubility was calculated via the following formula:

$$\text{Solubility } (\%) = \frac{\text{Protein content in supernatant}}{\text{Total protein content in sample}} \times 100 \quad (2)$$

2.13. Differential Scanning Calorimetry (DSC)

The thermal characteristics of spray-dried SPH and encapsulated SPH were evaluated by differential scanning calorimetry (DSC 214, Polyma, NETZSCH, Selb, Bavaria, Germany). After equilibration at 25 °C in desiccators over saturated salt solutions, aluminum pans with 3 mg of the sample were hermetically sealed. DSC analysis was performed at 10 °C/min from 30 to 350 °C in a nitrogen atmosphere. Indium was used as a standard for instrument calibration.

2.14. Fourier Transform Infrared (FTIR) Spectroscopy

The spectral profiles of spray-dried SPH and encapsulated SPH were analyzed via Fourier transform infrared spectroscopy (Alpha FTIR spectrometer, Bruker, Bremen, Germany) within the range of 4500–400 cm^{-1}. The reflectance spectra were converted to transmission spectra via the Kubelka–Munk algorithm for further interpretation.

2.15. Scanning Electron Microscopy (SEM)

The surface morphology of spray-dried SPH and encapsulated SPH was examined using a Sigma 300, Carl Zeiss scanning electron microscope (Oberkochen, Baden-Württemberg, Germany). A thin gold layer was applied to SEM stubs with carbon tabs to improve the conductivity. Accelerating voltages of 5 kV, 1.00K X and 3.00K X-ray micrographs were taken.

2.16. Bioactivity Property Analysis

2.16.1. DPPH (1,1-diphenyl-2-picrylhydrazyl) Radical-Scavenging Activity

The DPPH radical-scavenging activity of the protein hydrolysates was assessed by mixing the hydrolysate solutions (2, 4, 6, 8 and 10 mg/mL) with 0.1 mM DPPH. After a 30-min dark incubation, the absorbance was measured at 517 nm [44]. The DPPH scavenging activity (%) was calculated using the following formula:

$$\text{DPPH activity}(\%) = \frac{\text{Acontrol} - \text{Asample}}{\text{Acontrol}} \times 100 \quad (3)$$

where Acontrol is the absorbance of the DPPH solution without hydrolysate and Asample denotes the absorbance of the DPPH solution with hydrolysate.

2.16.2. Ferric-Reducing Antioxidant Power (FRAP) Assay

The ferric-reducing antioxidant power (FRAP) of the hydrolysates was assessed following the protocol of Pulido et al. [45] with some modifications. The FRAP reagent was prepared by mixing 25 mL of acetate buffer (pH 3.6), 2.5 mL of 10 mM TPTZ in 40 mM HCl, and 2.5 mL of 20 mM $FeCl_3 \cdot 6H_2O$. A 30 μL aliquot of hydrolysate was combined with 900 μL of the FRAP reagent and incubated at 37 °C for 30 min. The absorbance was then measured at 595 nm. Antioxidant activity was measured via a standard curve generated from iron (II) sulfate solutions and expressed as millimoles of ferrous equivalents per gram of hydrolysate (mM Fe^{2+}/g).

2.16.3. Hydroxyl Radical-Scavenging Activity (HRSA)

The hydroxyl radical-scavenging activity of the hydrolysates was evaluated using a modified method from You et al. [46]. The reaction mixture included 1.0 mL of phosphate-buffered saline (0.15 mol/L, pH 7.4), 1.0 mL of 1.0 mM safranin, 0.5 mL of 2.0 mmol/L EDTA-$FeSO_4$, and 1.0 mL of hydrolysate supernatant. After mixing, 1.0 mL of 3% H_2O_2 was added, and the mixture was incubated at 37 °C for 30 min. The absorbance was measured at 520 nm, and the scavenging activity (%) was calculated via the following formula:

$$\text{HRSA activity}\ (\%) = \frac{A_1 - A_0}{A_0} \times 100 \quad (4)$$

where A_1, A_2, and A_0 represent the absorbances of the supernatant, the mixture without H_2O_2, and the control, respectively. All the experiments were conducted in triplicate.

2.16.4. ABTS (2,2-azinobis (3-ethylbenzothiazoline-6-sulfonic acid)) Radical-Scavenging Activity

To evaluate the ABTS radical-scavenging activity, an ABTS solution was prepared by combining 7 mM ABTS with 2.45 mM potassium persulfate in equal parts. The solution was then stored in the dark for 16 h. The solution was diluted with distilled water, resulting in an absorbance of 0.70 ± 0.02 at 734 nm. To assess the scavenging ability, 40 μL of hydrolysate was combined with 4 mL of diluted ABTS solution. The mixture was then vortexed for 30 s and allowed to settle for 6 min in the dark. The absorbance was measured at 734 nm [47]. The ABTS radical-scavenging activity (%) was estimated via the following formula:

$$\text{ABTS activity}\ (\%) = \frac{\text{Acontrol} - \text{Asample}}{\text{Acontrol}} \times 100 \quad (5)$$

where Acontrol is the absorbance of the ABTS solution without the hydrolysate and Asample denotes the absorbance of the ABTS solution with the hydrolysate.

2.17. Statistical Analysis

The data from three replicates were statistically analyzed using analysis of variance (ANOVA). To identify significant differences among the means, Duncan's multiple range test was employed, with a significance threshold of 5% ($p < 0.05$). The statistical analysis was performed using SPSS software (version 23.0 for Windows; SPSS Inc., Chicago, IL, USA).

3. Results and Discussion

3.1. Degree of Hydrolysis (DH)

The extent of protein degradation by bromelain was quantified by determining the DH. The DH values measured at different time intervals are presented in Figure 3. After 300 min, the DH plateaued, indicating no further increase in hydrolysis. After 300 min of enzyme reaction, hydrolysates were made, with a DH of 48.05% from the soft tissue proteins of the snail. This study's DH value is slightly lower than the DH value reported by Auwal et al. [48] for stonefish protein hydrolysates made with bromelain after 360 min of hydrolysis. Different protein sources may have different amounts of DH because of changes in the structure and number of amino acids or because enzymes may not work properly [49,50]. According to Da Rosa Zavareze et al. [51], hydrolysates can be classified on the basis of DH into three categories, each with specific applications: low-DH hydrolysates improve protein functionality, medium-DH hydrolysates are generally used as flavor enhancers, and high-DH hydrolysates are suitable for nutritional supplements and specialized medical diets. Moreover, Vioque et al. [52] described hydrolyses with DH values above 10% as substantially hydrolyzed, whereas those with DH values less than 10% are regarded as somewhat hydrolyzed. With a DH of 48.05%, the hydrolysates developed in this work are thus highly hydrolyzed and can be used in food items to increase their bioactive qualities and increase their nutritional bioavailability.

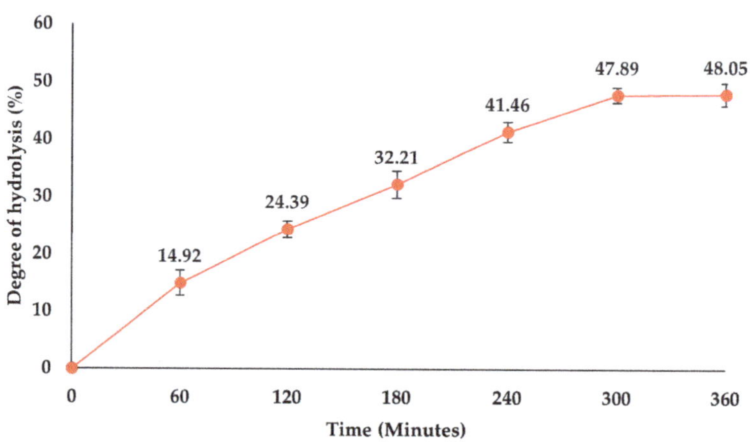

Figure 3. Degree of hydrolysis of snail protein hydrolysates at different time intervals. The values are presented as the mean ± SD ($n = 3$).

3.2. Yield

The yields of SPH and carrageenan-encapsulated SPH were assessed at varying concentrations of wall material (1%, 2%, and 3%). The findings, illustrated in Table 1, demonstrate a significant increase in yield from 19.74% at 1% wall material to 33.46% at 2% wall material ($p < 0.05$). A minor increase to 35.28% was noted for the 3% wall combination. The yield of SPH without encapsulation was 8.92%, which was substantially lower ($p < 0.05$) than that of carrageenan-encapsulated SPH. The notable increase in yield ($p < 0.05$) from 1% to

2% wall material indicates that carrageenan efficiently encapsulates and forms a protective matrix around the protein hydrolysates [53]. The marginal increase in yield at 3% may be attributed to the increased viscosity of the solution, which may impede the encapsulation process [54]. The lower yield of SPH relative to carrageenan-encapsulated SPH may be attributed to differences in protein structure and interactions with the encapsulating agent. Carrageenan's gelling properties may interact more effectively with the specific amino acid composition of SPH, enhancing encapsulation efficiency [55]. Additionally, SPH was more hygroscopic because of the higher content of charged N- or C-termini. During drying at high temperatures, greater evaporation of water occurred, reducing the moisture of the powders and increasing the capture of water molecules by the samples. When exposed to the environment, the dried powder rapidly assimilated moisture, causing the product to adhere inside the drying chamber [56]. In the presence of carrageenan, the particles were apparently larger than those constituting the soluble SPH.

Table 1. Physicochemical properties of snail protein hydrolysates (SPHs) and carrageenan-encapsulated SPH at different concentrations.

Parameters	SPH Without Encapsulation	1% Carrageenan Encapsulated SPH (CN 1%)	2% Carrageenan Encapsulated SPH (CN 2%)	3% Carrageenan Encapsulated SPH (CN 3%)
Yield (%)	8.92 ± 0.45^{a}	19.74 ± 0.65^{b}	33.46 ± 0.17^{c}	35.28 ± 0.61^{d}
EE (%)	-	73.31 ± 1.18^{a}	82.51 ± 0.10^{b}	84.96 ± 0.84^{c}
Hygroscopicity (%)	26.08 ± 0.40^{d}	16.88 ± 0.39^{c}	8.72 ± 0.64^{b}	7.24 ± 0.21^{a}
Lightness	89.33 ± 0.54^{a}	90.67 ± 0.81^{b}	92.54 ± 0.21^{c}	93.58 ± 0.30^{d}
Redness	0.93 ± 0.005^{d}	0.64 ± 0.015^{c}	0.38 ± 0.010^{b}	0.27 ± 0.005^{a}
Yellowness	13.78 ± 0.56^{d}	12.55 ± 0.38^{c}	11.46 ± 0.28^{b}	10.55 ± 0.30^{a}
Chroma	13.81 ± 0.56^{d}	12.57 ± 0.39^{c}	11.47 ± 0.28^{b}	10.56 ± 0.30^{a}
Whiteness	82.54 ± 0.77^{a}	84.33 ± 0.54^{b}	86.31 ± 0.23^{c}	87.64 ± 0.14^{d}

The values are presented as the mean ± SD (n = 3). Different superscripts indicate statistically significant differences between different groups.

3.3. Encapsulation Efficiency (EE)

Table 1 shows the encapsulation efficiency of carrageenan-encapsulated SPH at various wall material concentrations (1%, 2%, and 3%). The carrageenan concentration significantly improved the encapsulation efficiency ($p < 0.05$). The EE for the 1% wall material was 73.31%, which rose to 82.51% at 2% and 84.96% at 3%. These results indicate that increasing the proportion of carrageenan enhances the encapsulation efficiency, with the optimal efficiency achieved at 3%. Increased quantities of carrageenan enhance the encapsulation efficiency by forming a protective matrix surrounding protein hydrolysates. Carrageenan is recognized for its gelling and film-forming characteristics, which presumably enhance the stability and protection of SPH throughout the encapsulation process [57,58]. Recent studies on fish visceral protein hydrolysates indicate that increasing the concentration of wall materials improves encapsulation performance by forming a more cohesive barrier and minimizing the core material's exposure to outside factors such as moisture and heat [59,60]. However, while the encapsulation efficiency increased with increasing carrageenan concentration, the difference between the 2% and 3% carrageenan concentrations was negligible. These data suggest that the encapsulation efficiency may stagnate over a specific concentration of wall material. Excess carrageenan may also increase the viscosity of the solution, preventing adequate encapsulation and distribution of the protein hydrolysates within the matrix. This behavior has been confirmed in other encapsulation investigations, where excessively viscous fluids might reduce the efficiency of the encapsulation process [61,62].

3.4. Hygroscopicity

Table 1 lists the hygroscopicity of SPH and carrageenan-encapsulated SPH at different concentrations of wall material. The results demonstrated that SPH without encapsulation resulted in a substantially greater hygroscopicity of 26.08%, whereas encapsulation with carrageenan dramatically reduced hygroscopicity across all concentrations of wall material. The hygroscopicity at 1% carrageenan was 16.88%, which decreased to 8.72% at 2% carrageenan and reached a minimum of 7.24% at 3% carrageenan. These findings indicate that carrageenan encapsulation markedly ($p < 0.05$) diminishes the hygroscopicity of SPH, with the most substantial reduction noted at the maximum wall material concentration of 3%. The encapsulating method formed a barrier around the SPH particles that prevented them from coming into contact with outside moisture. This made them less hygroscopic when the concentration of carrageenan was high. The presence of carrageenan, a hydrocolloid known for its remarkable capacity to retain water and gel, probably aids in the formation of this protective matrix by decreasing the number of exposed protein sites that could interact with water molecules [63]. Comparable findings have been reported in fish oil encapsulated with maltodextrin and gum arabic, which demonstrated reduced hygroscopicity due to the formation of a stable matrix that shields hydrophilic peptide chains [64]. Similarly, encapsulation of chicken meat protein hydrolysates with maltodextrin and gum arabic resulted in a significant decrease in moisture absorption, attributed to the hydrocolloid's ability to form a dense, moisture-resistant coating [65]. The increased hygroscopicity observed in SPH without encapsulation can be attributed to the presence of charged amino acids and peptide chains that efficiently absorb moisture from the environment, resulting in increased water intake [66,67]. The significant reduction in hygroscopicity resulting from carrageenan encapsulation illustrates the effectiveness of this technique in improving the stability and shelf life of SPH, particularly in powdered formulations. An increase in wall material quantity (3%) may increase the viscosity of the encapsulation solution, perhaps leading to less uniform encapsulation. Nevertheless, the reduction in hygroscopicity at 3% indicates that carrageenan remains an effective barrier against moisture absorption. These findings correspond with prior studies that demonstrated the advantageous effects of carrageenan encapsulation in improving the functional properties of sensitive bioactive compounds [68].

3.5. Color Analysis

The color properties of the snail protein hydrolysate (SPH) and encapsulated SPH with varying concentrations of carrageenan (1%, 2%, and 3%) are presented in Table 1. The color properties of snail protein hydrolysate (SPH) and SPH encapsulated with carrageenan (at 1%, 2%, and 3% concentrations) showed notable variations that could improve its suitability for food applications. Encapsulation progressively increased the lightness (L^*) from 89.33 in nonencapsulated SPH to 93.58 in samples with 3% carrageenan, suggesting a brightening effect due to the carrageenan matrix [69]. Both the redness (a^*) and yellowness (b^*) values decreased with increasing carrageenan concentration, indicating that encapsulation diminishes the red and yellow tones, likely due to the masking effect of carrageenan [70]. The color intensity, measured by chroma (C^*), also decreased from 13.81 for the samples without encapsulation to 10.56 for the samples with 3% carrageenan encapsulation. This softening of color was accompanied by an increase in whiteness (W), which rose from 82.54 in nonencapsulated SPH to 87.64 with 3% carrageenan encapsulation, suggesting enhanced lightness and neutrality [71]. These findings show that encapsulated SPH may be advantageous in food products where a lighter and more neutral color is preferred, broadening its application potential.

3.6. Amino Acid Analysis

A total of 17 amino acids were identified in the SPH samples, as listed in Table 2, including seven essential amino acids (EAAs) (histidine, isoleucine, leucine, methionine, phenylalanine, tryptophan, and valine), two delicious amino acids (DAAs) (glutamine and proline), and four antioxidative amino acids (AAAs) (methionine, tyrosine, histidine and tryptophan). The findings revealed variable quantities of 17 amino acids, with histidine having the highest abundance (23.22%), followed by arginine (21.93%) and tyrosine (20.11%). These high quantities of basic and aromatic amino acids suggest a protein hydrolysate with potential nutritional value. Other important amino acids included proline (13.48%) and methionine (7.30%), whereas leucine (0.11%) and isoleucine (0.33%) were found at lower concentrations, indicating variation in essential amino acid distribution. The total percentage of hydrophobic amino acids was calculated to be approximately 24.31%, which is significant and suggests enhanced solubility and bioactivity, especially when these amino acids are included in antioxidant peptides or functional foods. Amino acids with antioxidant properties constitute approximately 52.54% of the total amino acids. This high antioxidative content reveals the resistance of the hydrolysate to oxidative stress, which is essential for cellular health and disease prevention [72]. The substantial presence of antioxidative amino acids, notably histidine (23.22%) and tyrosine (20.11%), suggests that *Bellamya bengalensis* protein hydrolysates might serve as valuable sources for the development of functional foods with antioxidant properties. Histidine is recognized for its capacity to buffer and chelate metal ions, potentially safeguarding cells from oxidative damage, which is vital in the prevention of aging and chronic diseases [72]. Tyrosine, another antioxidative amino acid, promotes neurotransmitter formation and may improve cognitive function, which contributes to the possible health advantages of this hydrolysate [73]. The presence of sulfur-containing amino acids, such as methionine (7.30%) and cysteine (1.91%), confirms their importance in protein synthesis and detoxification processes, both of which are critical for metabolic health [74]. The high abundance of hydrophobic amino acids (24.31%), particularly proline (13.48%) and methionine, was associated with increased solubility and bioactivity. Hydrophobic amino acids are frequently utilized in the synthesis of bioactive peptides, which can have antioxidant, antihypertensive, and antibacterial characteristics, making the hydrolysate appropriate for use in sports nutrition and medicinal supplements [75]. This is especially important considering the increasing global need for clean-label, bioavailable protein sources.

From a nutritional point of view, the essential amino acid content, which includes lysine, threonine, and methionine, suggests that it could be used as a supplementary protein source, especially in diets with low animal protein intake [76]. However, the low levels of essential amino acids, such as leucine and valine, in the hydrolysate may not be sufficient as a sole protein source. This limitation highlights the necessity of including other protein sources, such as plant-based proteins or dairy, to provide a balanced amino acid profile for human consumption that meets dietary requirements for essential amino acid intake [77]. The increased concentrations of tyrosine and histidine, together with the antioxidant properties of cysteine, increase the ability of hydrolysates to prevent lipid oxidation and microbial spoilage [78]. These results highlight the biofunctional properties of *B. bengalensis* protein hydrolysates. The presence of essential amino acids such as tyrosine, histidine, and cysteine make these hydrolysates appropriate for functional foods, nutraceuticals, and food preservation. Future studies should focus on isolating certain peptides and their use to optimize their potential.

Table 2. Amino acids identified in snail (*Bellamya bengalensis*) protein hydrolysates.

Amino Acids	Percentage (%)
Histidine **	23.22
Serine	1.07
Arginine	21.93
Glycine	1.11
Aspartic acid	0.94
Glutamic acid	0.35
Threonine	3.07
Alanine *	2.13
Proline *	13.48
Cysteine **	1.91
Lysine	1.97
Tyrosine **	20.11
Methionine *,**	7.30
Valine *	0.51
Isoleucine *	0.33
Leucine *	0.11
Phenylalanine *	0.45

* Hydrophobic amino acids, ** Antioxidative amino acids.

3.7. Particle Size Distribution and Polydispersity Index (PDI)

The particle size of protein hydrolysates (SPHs) affects their functional properties considerably, including their solubility, dispersion, and bioavailability. In this study, the particle size of spray-dried SPH was analyzed with and without carrageenan as a wall material, and the results are presented in Figure 4. Without encapsulation, the average size of SPH particles was 81.74 nm, suggesting enhanced solubility and bioavailability due to their larger surface area, which supports better dispersion and potential for functional ingredient applications [79,80]. When carrageenan was incorporated as a wall material at concentrations of 1%, 2%, and 3%, the particle sizes of the hydrolysate increased by 55% (127.09 nm), 129% (187.35 nm), and 153% (206.9 nm), respectively. This increase in the particle size confirms the coating effect on the hydrolysate. Furthermore, the increase in particle size may be attributed to the thickening and gelling properties of carrageenan, which promote greater particle formation during spray drying [81]. The polysaccharide likely forms a protective layer around SPH, resulting in more stable particles [82]. However, a larger particle size, while enhancing stability, may reduce solubility and limit SPH functionality in specific applications [83].

Despite the significant differences in particle size, the zeta potential values showed negligible variation across carrageenan concentrations. This is because zeta potential measures surface charge, not physical size. The encapsulated particles retained similar surface electrostatic behavior, indicating that colloidal stability was maintained regardless of particle size differences. Thus, larger sizes are due to matrix thickness, while stable zeta potential reflects consistent charge repulsion [82,84].

The polydispersity index (PDI) values offer insight into particle uniformity, with lower PDI values (<0.4) indicating more homogeneous distributions [85,86]. The PDI of unencapsulated SPH was 0.41, indicating a relatively broad particle size distribution, which may be due to particle aggregation from hydrophobic interactions [87,88]. Encapsulation with 1%, 2% and 3% carrageenan reduced the PDI to 0.23, 0.24 and 0.26, respectively, reflecting more consistent particle sizes and improved stability across all concentrations [89]. The lowest PDI value at 1% carrageenan was attributed to optimal stability at this concentration. However, a slight increase in the PDI at higher concentrations may result from minor particle size variations due to a thicker encapsulating layer [90]. A lower PDI enhances the functional stability of encapsulated bioactive peptides by promoting uniform particle behavior, which is beneficial for applications requiring controlled release [90,91].

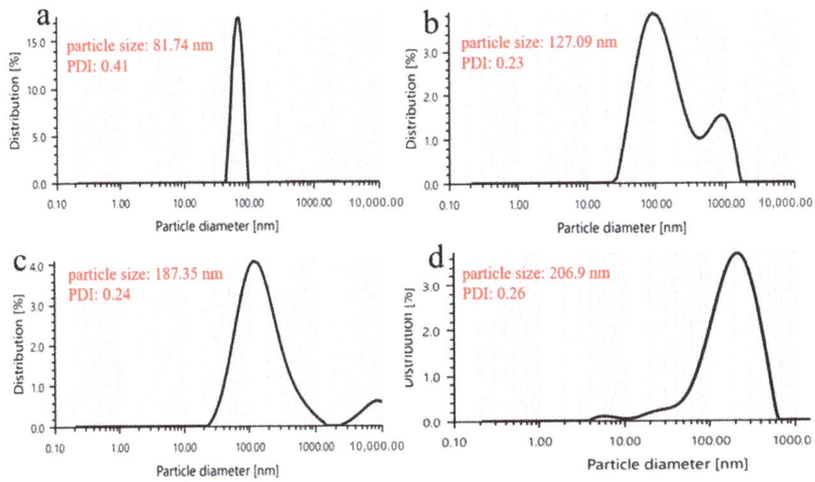

Figure 4. Particle size distributions of SPH and carrageenan-encapsulated SPH at different concentrations: (**a**) SPH without encapsulation, (**b**) 1% carrageenan-encapsulated SPH, (**c**) 2% carrageenan-encapsulated SPH, and (**d**) 3% carrageenan-encapsulated SPH.

3.8. Zeta Potential Measurements

The surface charge of spray-dried SPH and its encapsulated forms with varying concentrations of carrageenan was assessed by measuring the zeta potential, as shown in Figure 5. These parameters provide insights into the stability, particle interactions, and electrokinetic properties, which are critical for applications in food preservation and bioactive delivery systems [92]. The zeta potential of unencapsulated SPH was recorded at −11.8 mV, indicating considerable electrostatic repulsion among the particles. Nonetheless, the use of carrageenan resulted in a remarkable increase in the negative charge. The zeta potential values for SPH containing 1%, 2%, and 3% carrageenan were −25.3, −31.9, and −33.8 mV, respectively. The increase in the negative charge is attributed to the sulfated groups in carrageenan, which provide supplementary negative charges to the system [93]. The elevated negative zeta potential values indicate the enhanced colloidal stability of the encapsulated SPH formulations, which is essential for sustaining dispersion and preventing aggregation in food systems [94]. These findings indicate that carrageenan serves as an excellent encapsulating agent for protein hydrolysates, enhancing their applicability in food preservation and controlled release applications. Subsequent research may investigate the bioactivity and shelf durability of these encapsulated systems in real food matrices.

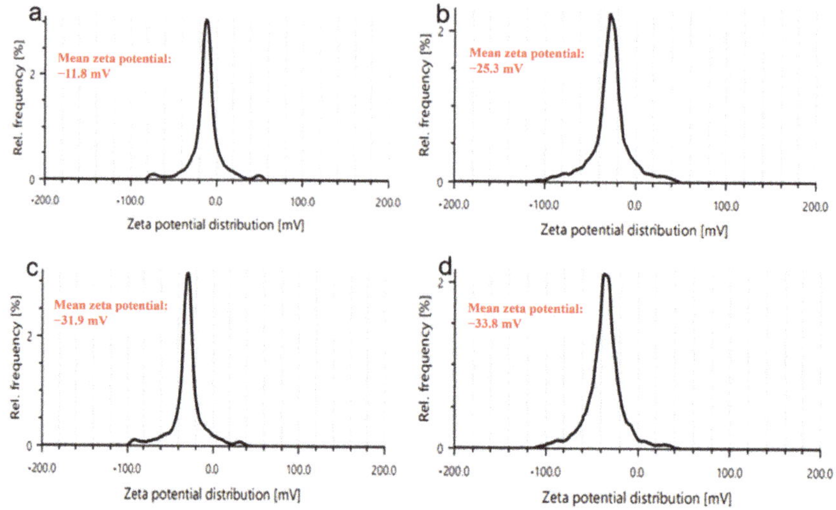

Figure 5. Zeta potential of SPH and carrageenan-encapsulated SPH: (**a**) SPH without encapsulation, (**b**) 1% carrageenan-encapsulated SPH, (**c**) 2% carrageenan-encapsulated SPH, (**d**) 3% carrageenan-encapsulated SPH.

3.9. Solubility

Table 3 presents the solubilities of spray-dried SPH and its encapsulated variants with carrageenan at various concentrations (1%, 2%, and 3%) throughout a range of pH values (2–10). Solubility is a major functional property of protein hydrolysates, considerably affecting their application in food and nutraceuticals [95]. The solubility of encapsulated hydrolysates is an essential indicator of their efficacy in delivering bioactive chemicals [96]. The solubility of nonencapsulated SPH was generally greater over the pH range, with minor fluctuations. At pH 2, the solubility was 97.32%, increasing to nearly complete solubility at pH 4 (99.97%) and sustaining a high level throughout the range, with a minor decrease to 94.61% at pH 10. These data demonstrate that SPH has exceptional solubility across both acidic and alkaline pH ranges. Encapsulation with carrageenan at different concentrations resulted in reduced solubility across all pH ranges. At 1% carrageenan, the solubility ranged from 72.13% at pH 2 to 70.87% at pH 10. With 2% carrageenan, the solubility ranged from 58.31% at pH 2 to 51.76% at pH 10. Finally, at 3% carrageenan, the solubility was the lowest, ranging from 43.28% at pH 2 to 41.10% at pH 10. The results indicate that the solubility of SPH is exceptionally high, particularly in acidic environments (pH 2–4), which is consistent with prior findings indicating that protein hydrolysates generally exhibit good solubility because of their relatively small peptide sizes and the presence of polar amino acid residues [19,97]. This high solubility across a broad pH range makes SPH a suitable ingredient for inclusion in various food matrices, where pH conditions may vary.

The encapsulation of SPH with carrageenan significantly reduced the solubility, with elevated concentrations of carrageenan (CN 3%) yielding the most substantial reduction. At pH 2, the solubility decreased from 97.32% for SPH to 43.28% for CN 3%. The reduction in solubility is probably attributable to the interaction between the carrageenan matrix and the protein hydrolysates, which may restrict the exposure of the peptides to the aqueous environment and decrease their solubility [98]. Despite this reduction, encapsulation could still be beneficial, mainly in applications where controlled release or protection of bioactive peptides is desired. A lower carrageenan concentration (1%) resulted in better solubility, suggesting that a balance between encapsulation and maintaining functional properties such as solubility can be achieved by optimizing the carrageenan content. However, the decreased solubility at higher carrageenan concentrations may limit the effectiveness of these encapsulated hydrolysates in certain food applications, especially in products requiring high solubility.

Table 3. Solubility (%) of SPH and carrageenan-encapsulated SPH at different pH values.

	Solubility (%)				
pH	2	4	6	8	10
Snail protein hydrolysates	97.31 ± 0.81 ^{bD}	99.96 ± 0.020 ^{dD}	99.26 ± 0.18 ^{cdD}	98.35 ± 0.48 ^{cD}	94.61 ± 0.65 ^{aD}
Carrageenan-encapsulated SPH (1%)	72.13 ± 0.40 ^{bC}	79.83 ± 0.58 ^{eC}	77.84 ± 0.65 ^{dC}	74.10 ± 0.98 ^{cC}	70.87 ± 0.42 ^{aC}
Carrageenan-encapsulated SPH (2%)	58.31 ± 0.90 ^{bB}	63.32 ± 0.96 ^{dB}	60.39 ± 0.55 ^{cB}	57.06 ± 0.74 ^{bB}	51.75 ± 0.84 ^{aB}
Carrageenan-encapsulated SPH (3%)	43.28 ± 0.56 ^{bA}	45.73 ± 0.63 ^{cA}	46.89 ± 0.76 ^{cA}	43.77 ± 0.62 ^{bA}	41.10 ± 0.78 ^{aA}

The values are presented as the mean ± SD (n = 3). Different superscripts between pH values (lowercase letters) and groups (uppercase letters) indicate statistically significant differences.

3.10. Differential Scanning Calorimetry (DSC)

The DSC thermograms for spray-dried SPH and its encapsulated forms with 1%, 2%, and 3% carrageenan are shown in Figure 6. DSC analysis provides insight into the thermal transitions and stability of samples, which is essential for understanding their encapsulation effects and potential applications in food preservation [99]. The DSC thermogram of SPH without encapsulation shows an endothermic peak near 60 °C, indicating the denaturation or thermal transition of protein components [100]. This is a common feature in protein hydrolysates due to the unfolding of peptide chains when subjected to heat. Beyond this temperature, the DSC curve gradually increases to approximately 325 °C, indicating the thermal degradation of the protein structure, which is characteristic of the breakdown of organic molecules into smaller volatile compounds [101].

The thermal profile with 1% carrageenan encapsulation showed a minor alteration in the initial endothermic transition at 65 °C. This transition indicates improved thermal stability attributable to the carrageenan matrix, which protects the protein hydrolysates. The widening of the transition indicates enhanced contact between the protein and the encapsulating agent, probably via hydrogen bonding or electrostatic interactions [102]. The delayed initiation of thermal degradation indicates enhanced resistance to heat-induced denaturation. The 2% carrageenan-encapsulated SPH sample exhibited a significant thermal transition at approximately 70 °C, indicating the stabilizing influence of the increased carrageenan concentration. An elevated concentration of carrageenan is expected to improve the encapsulation efficiency, offering enhanced protection against thermal denaturation [103]. A sharper peak indicates a more distinct thermal transition, perhaps resulting from the enhanced development of a protective matrix surrounding the protein hydrolysates. Above 250 °C, the degradation slope is less pronounced than that of the unencapsulated SPH, indicating enhanced thermal stability. The 3% carrageenan-encapsulated SPH mixture has a similar DSC curve with a stronger thermal transition peak at approximately 75 °C, indicating that it is the most thermally stable sample of the encapsulated formulations. The substantial increase in the endothermic peak signifies that the encapsulating matrix at this concentration is particularly efficient in stabilizing the protein hydrolysates against thermal degradation. Moreover, the initiation of thermal degradation was postponed relative to that of the lower-concentration samples, indicating that elevated carrageenan concentrations offer improved protection. Moreover, the extensive degradation profile above 300 °C indicates that carrageenan at elevated concentrations may undergo thermal degradation, thus affecting the overall thermal dynamics of the system [101].

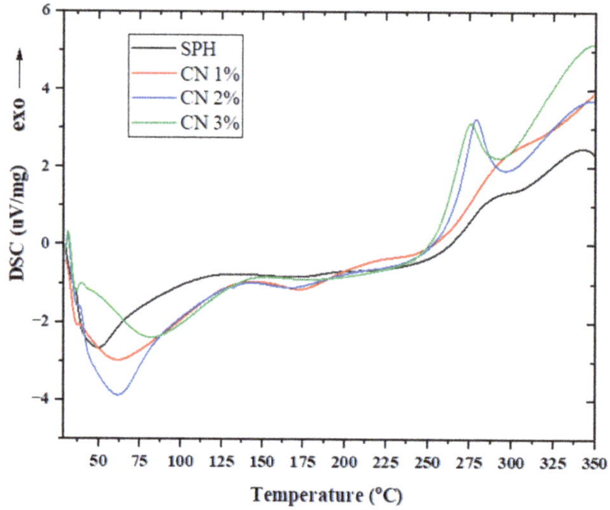

Figure 6. Differential scanning calorimetry curves of SPH and carrageenan-encapsulated SPH at different concentrations. The upward arrow indicating the direction of exothermic transitions.

3.11. Fourier Transform Infrared (FTIR) Spectroscopy

FTIR spectroscopy was used to evaluate the structural interactions and molecular characteristics of the encapsulated SPH at various concentrations (1%, 2%, and 3%) of carrageenan (CN). The spectra displayed in Figure 7 and Table 4 show significant peaks corresponding to various functional groups, which provide insights into the encapsulation process and the interaction between the protein hydrolysates and the encapsulating agent.

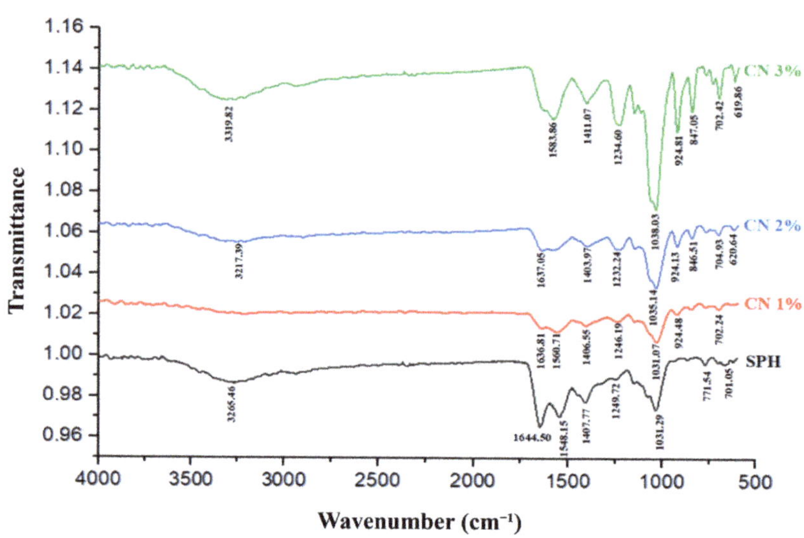

Figure 7. FTIR spectra of SPH and carrageenan-encapsulated SPH at different concentrations.

Table 4. FTIR analysis of SPH and carrageenan-encapsulated SPH.

Band Position (cm^{-1})	Functional Group/Vibration Type	SPH Without Encapsulation	1% Carrageenan-Encapsulated SPH	2% Carrageenan-Encapsulated SPH	3% Carrageenan-Encapsulated SPH
3319.92–3265.46	O–H and N–H stretching (hydrogen bonding)	3265.46	3217.39	3285.15	3319.92
1583.86–1644.50	Amide I band (C=O stretching of peptide bonds)	1644.50	1635.81	1637.05	1583.86
1548.15–1411.07	Amide II band (N–H bending, C–N stretching)	1548.15	1560.71	1403.97	1411.07
1407.77–1403.97	CH_2 bending vibration	1407.77	1506.57	1403.97	Nil
1249.72–1223.24	Amide III band (N–H bending, C–N stretching)	1249.72	1246.60	1223.24	1234.60
1038.63–1031.29	C–O stretching (carbohydrate/protein side chains)	1031.29	1036.14	1038.63	1038.63
924.81–924.48	C–O–S or C–O–C stretching (polysaccharide/carrageenan)	Nil	924.48	924.45	924.81
847.05	Sulfate group (S=O symmetric stretching)	Nil	Nil	Nil	847.05
702.42–702.24	Sulfate group (S=O stretching in carrageenan)	701.05	702.24	702.64	702.42
771.54–771.45	C–H out-of-plane bending or secondary protein structures	771.45	Nil	Nil	Nil
620.64–619.88	Sulfate group (possible S–O vibration)	Nil	Nil	620.64	619.88

The FTIR spectra of SPH (black line) exhibited significant absorption bands at 3265.46 cm^{-1}, 1644.50 cm^{-1}, 1548.15 cm^{-1}, 1407.77 cm^{-1}, 1249.72 cm^{-1}, 1031.29 cm^{-1}, and 771.45 cm^{-1}. The prominent peak at approximately 3265.46 cm^{-1} is attributed to the stretching vibrations of O–H and N–H bonds, indicating the presence of free hydroxyl and amine groups, which are characteristic of proteinaceous materials [104]. The amide I and amide II bands are detected at 1644.50 cm^{-1} and 1548.15 cm^{-1}, respectively, corresponding to the C=O stretching and N-H bending vibrations of peptide bonds, confirming the presence of α-helices and β-sheets [105,106]. The peak at 1407.77 cm^{-1} is attributed to CH_2 bending, whereas the peak at 1249.72 cm^{-1} is attributed to the amide III band, which is connected with N–H bending and C–N stretching. The peaks at 1031.29 cm^{-1} and 771.45 cm^{-1} signify C–O stretching and other secondary structures in the protein hydrolysates [107]. In the FTIR spectrum of samples encapsulated with 1% carrageenan (red line), the characteristic peaks of SPH are retained but show noticeable shifts, indicating interactions between SPH and carrageenan. The O–H and N–H stretching peaks slightly shifted to 3217.39 cm^{-1}, indicating hydrogen bonding interactions between the protein hydrolysate and carrageenan. The amide I and amide II bands are observed at 1635.81 cm^{-1} and 1560.71 cm^{-1}, showing a minor shift from the SPH spectrum, which suggests some encapsulation effects on the protein's secondary structure [108]. Another peak at 702.24 cm^{-1} is attributed to the sulfate group (S=O stretching) of carrageenan, confirming its incorporation into the matrix [109]. For 2% CN (blue line), the FTIR spectrum shows further shifts in the key absorption bands. The O–H/N–H stretching band appears at 3285.15 cm^{-1}, which is slightly similar to that of CN 1%, indicating continued hydrogen bonding between carrageenan and SPH. The amide I and amide II bands are observed at 1637.05 cm^{-1} and 1403.97 cm^{-1}, suggesting that the secondary structure of SPH [108] is further affected by the increased carrageenan concentration. A notable peak at 620.64 cm^{-1} was observed, which may also be associated with the sulfate groups of carrageenan [109].

The additional sulfate peak and the shifting of the amide bands confirmed the increased interaction between carrageenan and the protein hydrolysates at this concentration. The FTIR spectrum of CN 3% (green line) shows more pronounced shifts in the O–H/N–H stretching region, with a peak at 3319.92 cm^{-1}, which is a significant shift compared with both the SPH and the lower carrageenan concentrations. This suggests stronger hydrogen bonding and more extensive encapsulation interactions at the highest concentration of carrageenan. The amide I and II bands shifted to 1583.86 cm^{-1} and 1411.07 cm^{-1}, respectively, indicating more significant conformational changes in the protein hydrolysate secondary structure due to encapsulation [110]. The peaks at 924.81 cm^{-1}, 847.05 cm^{-1}, and 702.42 cm^{-1} further confirmed the presence of carrageenan, with the sulfate group peaks becoming more pronounced as the concentration increased [109].

FTIR examination validated the effective encapsulation of SPH by carrageenan, as shown by shifts in the O–H/N–H and amide I/II bands, signifying hydrogen bonding and structural alterations. Elevated CN concentration exacerbated these changes and introduced sulphate group peaks, underscoring enhanced interactions. The 3% CN formulation exhibited the most significant spectrum alterations, indicating improved encapsulation efficiency. These findings endorse the efficacy of carrageenan as a viable encapsulating agent for protein hydrolysates in functional food applications.

3.12. Scanning Electron Microscopy (SEM) Analysis

SEM analysis of spray-dried SPH, both unencapsulated and encapsulated with 1%, 2%, and 3% carrageenan, as shown in Figure 8, revealed a significant variation in particle morphology, providing insights into the efficiency and characteristics of the encapsulation process. For the unencapsulated SPH particles, SEM images at 1.0 KX and 3.0 KX magnifications reveal a heterogeneous surface morphology, with a mix of spherical and irregularly shaped particles ranging from 2 to 10 microns. These spherical particles typically exhibit smooth surfaces, although some surface irregularities, cracks, and pores are apparent, likely due to the spray-drying process [111]. These morphological characteristics align with previous findings on spray drying, which reported atomization-driven formation of spherical particles with variable sizes [112]. Surface irregularities could indicate structural changes during drying, potentially impacting solubility and functional properties [113]. In contrast, encapsulated SPH with 1% carrageenan resulted in smoother and more uniform spherical particles, with fewer surface irregularities, suggesting successful encapsulation that formed a cohesive carrageenan matrix around the hydrolysates. Small pores visible at higher magnification may facilitate controlled release, supporting the effectiveness of carrageenan as an encapsulation material for enhancing stability and protecting bioactive compounds [114,115]. Increasing the carrageenan concentration to 2% and 3% further impacted the particle morphology, with SEM images at 1.0 KX and 3.0 KX showing denser, more compact particles, reduced porosity and smoother surfaces. This enhanced uniformity and stability likely contribute to prolonged release and increased encapsulation efficiency, which are crucial for applications in functional foods and nutraceuticals requiring sustained release [116,117].

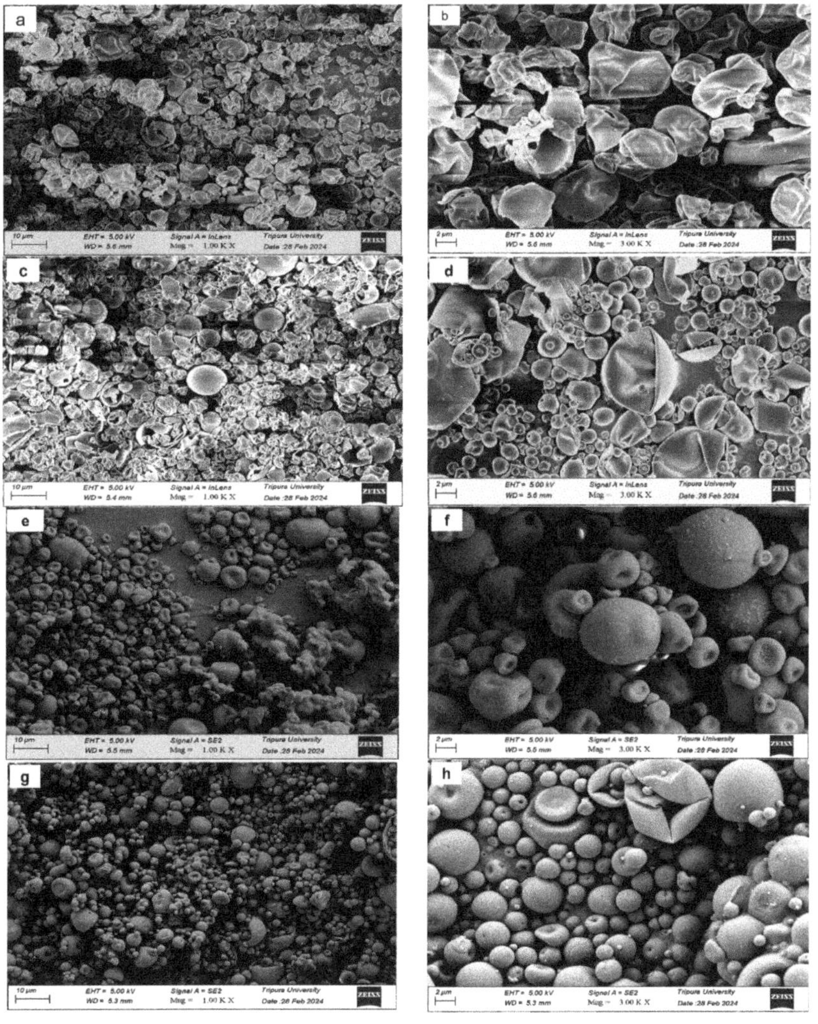

Figure 8. Scanning electron microscopy (SEM) images of SPH and carrageenan-encapsulated SPH at different concentrations and magnifications: (**a**) SPH without encapsulation at 1 KX magnification; (**b**) SPH without encapsulation at 3 KX magnification; (**c**) 1% carrageenan-encapsulated SPH at 1 KX magnification; (**d**) 1% carrageenan-encapsulated SPH at 3 KX magnification; (**e**) 2% carrageenan-encapsulated SPH at 1 KX magnification; (**f**) 2% carrageenan-encapsulated SPH at 3KX magnification; (**g**) 3% carrageenan-encapsulated SPH at 1 KX magnification; (**h**) 3% carrageenan-encapsulated SPH at 3 KX magnification.

3.13. Bioactivity Property Analysis

3.13.1. DPPH (1,1-diphenyl-2-picrylhydrazyl) Radical-Scavenging Activity

Figure 9 shows the DPPH radical-scavenging activity of SPH and encapsulated SPH using carrageenan at three different concentrations (1%, 2%, and 3%). The DPPH radical-scavenging activity of SPH and carrageenan-encapsulated SPH clearly increased in a dose-dependent manner across all the treatments, highlighting the strong antioxidant potential of SPH. For nonencapsulated SPH, the scavenging activity increased steadily from 54.64% at 2 mg/mL to 94.11% at 10 mg/mL, corroborating findings from previous studies on Skipjack tuna and Nile tilapia protein hydrolysates which exhibit increased radical scavenging capacity with increasing concentration [118,119]. The dose-dependent nature of antioxidant activity can be attributed to the increased availability of peptides that donate protons or electrons to neutralize free radicals as the concentration increases [120].

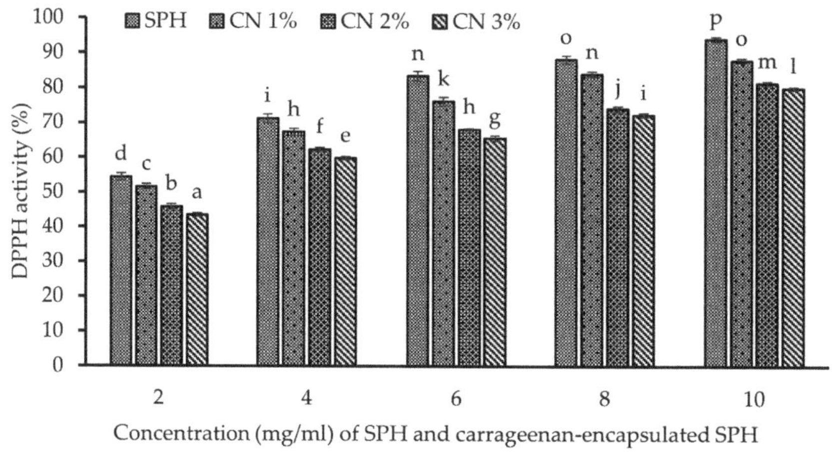

Figure 9. DPPH activity of SPH and carrageenan-encapsulated SPH at different concentrations. The data are expressed as the mean ± SD (n = 3). Bars with different superscripts are significantly different ($p < 0.05$).

SPH encapsulated with carrageenan also exhibited significant radical-scavenging activity, although it was slightly lower than that of nonencapsulated SPH. At 10 mg/mL, the scavenging activity for 1% carrageenan (CN 1%) was 88.46%, that for 2% carrageenan (CN 2%) was 81.53%, and that for 3% carrageenan (CN 3%) was 80.03%, suggesting that encapsulation at higher carrageenan concentrations modestly reduces the antioxidant efficacy. This decrease is due to the interaction between carrageenan and the bioactive peptides in SPH, since polysaccharides such as carrageenan can partially encapsulate or protect the active regions of peptides, limiting their accessibility to DPPH radicals [121]. Furthermore, prior studies have demonstrated that encapsulation with a relatively high carrageenan concentration slightly slows the release of bioactive chemicals by forming a relatively complex, dense matrix [122]. However, the encapsulated SPH still exhibited significant scavenging activity, indicating that encapsulation does not significantly reduce bioactivity.

Despite the minor reduction in antioxidant efficiency, the encapsulation technique has significant advantages. Encapsulation with carrageenan may increase the stability of SPH by preserving the peptides from oxidation and degradation, making it more useful in applications requiring long-term antioxidant action [91]. Carrageenan, a natural polysaccharide, has also been found to improve the stability and bioavailability of active compounds, thereby supporting the functional use of encapsulated SPH in food preservation or nutraceutical formulations [69].

3.13.2. Ferric-Reducing Antioxidant Power (FRAP) Assay

Figure 10 depicts the ferric-reducing antioxidant power activity of SPH and its encapsulated forms at concentrations of 1%, 2%, and 3% carrageenan. The concentration of nonencapsulated SPH was substantially related to the FRAP activity, with values ranging from 57.08 mM Fe(II)/g at 2 mg/mL to 95.26 mM Fe(II)/g at 10 mg/mL. This pattern is consistent with prior studies, which determined that the antioxidant potential of protein hydrolysates from common carp and ray finned fish increases with increasing concentrations due to the presence of peptides that serve as electron donors, converting ferric ions to ferrous forms [123]. The linear increase in FRAP values suggests that SPH has strong electron-donating capabilities, most likely due to the bioactive peptides produced during hydrolysis, which have been demonstrated to significantly contribute to reducing power and consequently antioxidant capability [124,125]. High FRAP values are associated with the presence of particular amino acids, such as valine, leucine, proline, histidine, and tyrosine, which play important roles in increasing antioxidant activity when combined into peptide sequences. Among these, tyrosine is an excellent electron donor that efficiently captures electrons to prevent the oxidation of other molecules by reactive chemicals [126,127].

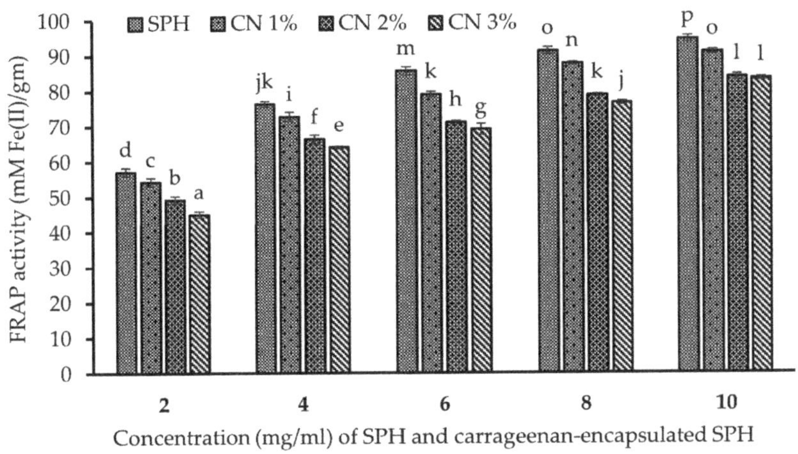

Figure 10. FRAP activity of SPH and carrageenan-encapsulated SPH at different concentrations. The data are expressed as the mean ± SD ($n = 3$). Bars with different superscripts are significantly different ($p < 0.05$).

SPH encapsulated with carrageenan at 1%, 2%, and 3% concentrations showed a steady increase in FRAP activity as the SPH concentration increased, while the values were slightly lower than those of nonencapsulated SPH across all carrageenan concentrations. At 10 mg/mL SPH, the FRAP activity was 91.39 mM Fe(II)/g for 1% carrageenan, 83.84 mM Fe(II)/g for 2% carrageenan, and 83.47 mM Fe(II)/g for 3% carrageenan. This slight reduction in FRAP activity for the encapsulated samples could be attributed to the presence of carrageenan, which can partially inhibit the interaction between SPH peptides and ferric ions [69]. As the concentration of carrageenan increases, the encapsulation matrix may inhibit peptide diffusion, restricting the ability of the peptides to act as electron donors in the ferric ion reduction process [68]. Nasri et al. [1] reported that at higher biopolymer concentrations, the encapsulating matrix can become denser, limiting the diffusion of antioxidant molecules from within the encapsulation. This reduced release can lead to a decrease in the antioxidant's immediate availability and overall activity, as the bioactive compounds are less accessible for interaction with target molecules.

3.13.3. Hydroxyl Radical-Scavenging Activity

Figure 11 shows the hydroxyl radical-scavenging activity of SPH and its encapsulated forms with carrageenan at different concentrations. The results of the hydroxyl radical-scavenging activity of SPH and its carrageenan-encapsulated derivatives indicated a positive relationship between SPH concentration and antioxidant potential. Hydroxyl radicals are the most reactive oxygen species, producing significant oxidative damage by damaging biomolecules such as lipids, proteins, and DNA [128,129]. The capacity of these protein hydrolysates to neutralize hydroxyl radicals suggests a potential role in reducing oxidative stress and its negative effects [130,131]. The nonencapsulated SPH exhibited a significant increase in scavenging activity with increasing SPH concentration, ranging from 28.37% at 2 mg/mL to 64.61% at 10 mg/mL. This trend is consistent with previous studies indicating that higher concentrations of protein hydrolysates improve antioxidant capacity, likely due to an increased number of bioactive peptides capable of interacting with and neutralizing reactive oxygen species (ROS) [11,132]. Bioactive peptides, particularly those containing amino acids such as tyrosine, histidine, and cysteine, have been shown to act as hydrogen donors, thereby stabilizing free radicals and reducing oxidative damage [133].

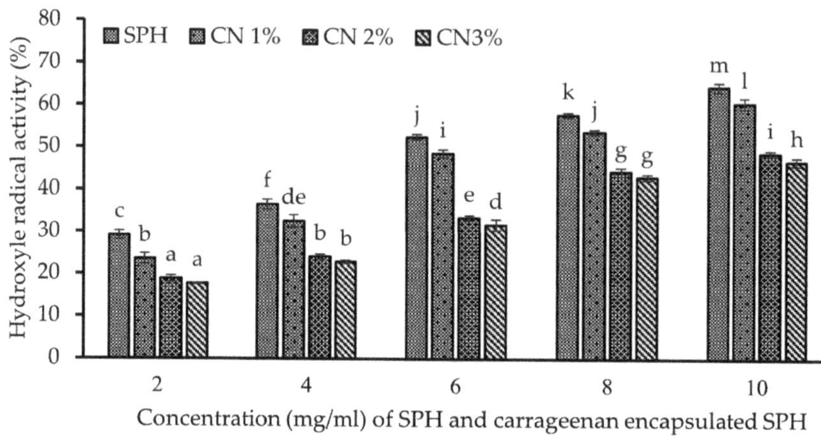

Figure 11. Hydroxyl radical-scavenging activity of SPH and carrageenan-encapsulated SPH at different concentrations. The data are expressed as the mean ± SD (n = 3). Bars with different superscripts are significantly different ($p < 0.05$).

The encapsulated SPH also tended to increase in hydroxyl radical-scavenging activity as the SPH concentration increased across all carrageenan concentrations (1%, 2%, and 3%). However, the encapsulated forms generally exhibited lower activity than nonencapsulated SPH did, likely because the encapsulation matrix partially restricted the release and accessibility of the active peptides [98]. At 1% carrageenan, the scavenging activity increased from 23.16% at 2 mg/mL to 60.67% at 10 mg/mL, whereas at 2% and 3% carrageenan, the scavenging activity ranged from 18.79% to 48.7% and 17.62% to 46.83%, respectively. This reduction in scavenging efficiency at higher carrageenan concentrations is likely due to the denser biopolymer matrix, which can impede the release and diffusion of peptides, thereby limiting their interaction with hydroxyl radicals [1]. Higher carrageenan concentrations increase the matrix density, thereby reducing the bioavailability of peptides within the hydrolysates, as observed in other biopolymer encapsulation systems [134]. Polysaccharide-based encapsulation has been shown to protect peptides from degradation, thereby increasing their functional longevity. Thus, while encapsulation might reduce immediate antioxidant activity, it provides potential advantages for long-term applications in food preservation or nutraceutical formulations, where a controlled release of antioxidants is beneficial [135].

3.13.4. ABTS (2,2-azinobis (3-ethylbenzothiazoline-6-sulfonic acid)) Activity

Figure 12 shows the ABTS radical-scavenging activity of SPH and its encapsulated forms at various carrageenan concentrations (1%, 2%, and 3%). The ABTS radical-scavenging activity quantifies the antioxidant capability, with elevated percentages indicating enhanced scavenging of ABTS radicals [136]. The ABTS radical-scavenging activity of nonencapsulated and encapsulated SPH increased from 2 mg/mL to 10 mg/mL. At 2 mg/mL, the nonencapsulated SPH demonstrated 43.42% scavenging activity, which progressively increased to 93.92% at 10 mg/mL (Figure 1). This increase indicates that the antioxidant capacity of SPH is dependent on concentration, which is consistent with findings from similar studies where elevated concentrations of protein hydrolysates led to increased radical-scavenging activity [123]. The results demonstrate that SPH comprises peptides that may donate electrons, enabling them to interact with free radicals and transform them into more stable molecules, effectively ending the radical chain reaction [137]. Compared with the nonencapsulated SPH process, the encapsulation process led to slightly reduced activity. For samples encapsulated with 1% carrageenan, the ABTS scavenging activity ranged from 41.67% at 2 mg/mL to 90.09% at 10 mg/mL, demonstrating relatively less activity than SPH without encapsulation. Similarly, SPH encapsulated with 2% carrageenan revealed a decrease in scavenging activity from 34.52% at 2 mg/mL to 81.51% at 10 mg/mL, whereas samples encapsulated with 3% carrageenan demonstrated an activity range of 31.96% to 79.55%. These data indicate that while encapsulation did not fully suppress the antioxidant potential of SPH, it lowered the overall scavenging capacity, which may be attributed to the physical barriers established by the carrageenan matrix, which restricts the accessibility of SPH to ABTS radicals [1,69]. The encapsulation of bioactive compounds is frequently employed to protect sensitive compounds, regulate their release, and improve their stability. However, the results of the present study suggest that their functional properties, including their antioxidant activity, may be marginally limited. The decrease in activity could result from the reduced solubility or mobility of the encapsulated protein hydrolysates within the carrageenan matrix. The encapsulated SPH demonstrated considerable antioxidant ability, making it a potential candidate for use in controlled delivery systems for food preservation or therapeutic applications.

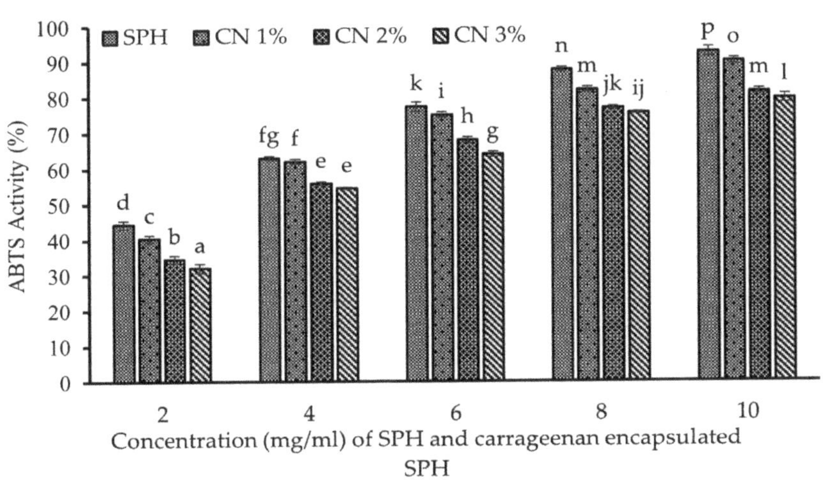

Figure 12. ABTS activity of SPH and carrageenan-encapsulated SPH at different concentrations. The data are expressed as the mean ± SD (n = 3). Bars with different superscripts are significantly different ($p < 0.05$).

Although the in vitro antioxidant activity of encapsulated SPH is encouraging, its practical effectiveness in food systems or gastrointestinal environments requires more investigation. Food matrices may affect the release kinetics and interaction of SPH with reactive oxygen species, thereby altering bioactivity [138]. Furthermore, simulated digestion models indicate that encapsulated peptides can demonstrate delayed yet prolonged antioxidant action during gastrointestinal transit due to regulated release mechanisms [17,139]. Consequently, forthcoming in vivo or simulated gastrointestinal investigations are essential to confirm the functional applicability of encapsulated SPH in authentic dietary and physiological contexts.

4. Conclusions

The protein hydrolysates prepared from *Bellamya bengalensis* exhibited increased yield, stability, and antioxidant potential through microencapsulation assisted by carrageenan. Encapsulation enhances structural integrity and reduces hygroscopicity, a major factor for maintaining bioactivity and functionality in food applications. Increasing the carrageenan concentration to 3% substantially stabilized the encapsulated SPH, as evidenced by larger particle sizes, reduced polydispersity, and increased zeta potential, contributing to better shelf stability. Although the antioxidant activity slightly decreased with increasing concentrations of carrageenan, significant bioactivity retention by SPH encapsulated in this polysaccharide shows great promise for controlled-release applications in food preservation and nutraceuticals. Our findings indicate that carrageenan is an excellent natural microencapsulating agent that enhances the stability and functional profile of snail protein hydrolysates, thus enabling their wider application as a bioactive ingredient within different food technologies.

While carrageenan effectively enhanced the stability and functionality of SPH, its practical applicability may be restricted by a modest loss in antioxidant activity at higher concentration. However, the longer shelf life, structural protection, and ability for controlled release make carrageenan-encapsulated SPH a viable component for functional foods and nutraceuticals. Further studies optimizing encapsulation conditions could help maximize both stability and bioactivity.

Author Contributions: Writing—original draft preparation, investigation, and formal analysis, A.V.; conceptualization, manuscript editing, supervision, and visualization, N.K.M.; writing—review and editing and supervision, M.B.P.; writing—review and editing, S.K.S.; formal analysis, P.C.A., S.B., H.N. and S.A.H.; data curation, P.P.; investigation, J.L., N.S.S. and B.K.P. All authors have read and agreed to the published version of the manuscript.

Funding: This research received no external funding.

Institutional Review Board Statement: The research conducted in this study involved the use of freshwater snails, which are commonly consumed as food in the northeastern regions of India. As these animals are considered edible and are regularly harvested for human consumption, no specific ethical permission was required. The study adhered to all applicable local and institutional guidelines for the ethical use of biological materials.

Informed Consent Statement: Not applicable.

Data Availability Statement: The data that support the findings of this study are available from the corresponding author upon reasonable request.

Acknowledgments: The authors are thankful to the Dean of the College of Fisheries, Central Agricultural University (Imphal), Lembucherra, Tripura, India, for providing the facilities necessary to carry out this research. The authors are also greatly thankful for the chemical support received from the Department of Biotechnology, Government of India, under the Centre of Excellence project (DBT-NER/LIVS/05/2011 Phase II), Central Agricultural University, Imphal, Manipur, India. The authors also acknowledge the Drug Metabolomics Laboratory, Department of Pharmacy, Tripura University (A Central University), for providing the necessary facilities and support to carry out this study.

Conflicts of Interest: The authors declare no conflicts of interest.

References

1. Nasri, R.; Hamdi, M.; Touir, S.; Li, S.; Karra-Chaâbouni, M.; Nasri, M. Development of delivery system based on marine chitosan: Encapsulation and release kinetic study of antioxidant peptides from chitosan microparticle. *Int. J. Biol. Macromol.* **2021**, *167*, 1445–1451. [CrossRef] [PubMed]
2. Senadheera, T.R.; Hossain, A.; Shahidi, F. Marine bioactives and their application in the food industry: A review. *Appl. Sci.* **2023**, *13*, 12088. [CrossRef]
3. Vásquez, P.; Cian, R.E.; Drago, S.R. Marine Bioactive Peptides (Fishes, Algae, Cephalopods, Molluscs, and Crustaceans). In *Handbook of Food Bioactive Ingredients*; Jafari, S.M., Rashidinejad, A., Simal-Gandara, J., Eds.; Springer: Cham, Switzerland, 2023; pp. 1–16. [CrossRef]
4. Bar, A. *Bellamya bengalensis*: A review on its Ecological importance, Nutritional values & Ethnomedicinal importance. *Eur. J. Pharm. Med. Res.* **2020**, *7*, 315–319.
5. Chakraborty, D.; Mukherjee, M.; Maity, J. Ethnomedicinal importance of viviparous gastropod, *Bellamya bengalensis* (Lamarck, 1822) in terms of estimation of its proximate protein and amino acid composition. *Int. J. Sci. Res.* **2015**, *3*, 245–248.
6. Gupta, A.; Khanal, P. The potential of snails as a source of food and feed. *J. Agric. Food Res.* **2024**, *18*, 101330. [CrossRef]
7. Pissia, M.A.; Matsakidou, A.; Kiosseoglou, V. Raw materials from snails for food preparation. *Future Foods* **2021**, *3*, 100034. [CrossRef]
8. Tacias-Pascacio, V.G.; Castañeda-Valbuena, D.; Morellon-Sterling, R.; Tavano, O.; Berenguer-Murcia, Á.; Vela-Gutiérrez, G.; Fernandez-Lafuente, R. Bioactive peptides from fisheries residues: A review of use of papain in proteolysis reactions. *Int. J. Biol. Macromol.* **2021**, *184*, 415–428. [CrossRef]
9. Khalua, R.K.; Tripathy, S.; Paul, B.; Bairy, D. Seasonal variation of carbohydrate, protein and lipid of common freshwater edible gastropod (*Bellamya bengalensis*) of Medinipur district, West Bengal. *Res. J. Biol.* **2014**, *2*, 49–52.
10. Debnath, C.; Sahoo, L.; Haldar, A.; Datta, M.; Yadav, G.S.; Singha, A.; Bhattacharjee, J. Proximate and mineral composition of freshwater snails of Tripura, North–East India. *Fish. Technol.* **2016**, *53*, 307–312.
11. Chatterjee, R.; Dey, T.K.; Roychoudhury, A.; Paul, D.; Dhar, P. Enzymatically excised oligopeptides from *Bellamya bengalensis* show potent antioxidative and anti-hypertensive activity. *J. Food Sci. Technol.* **2020**, *57*, 2586–2601. [CrossRef] [PubMed]
12. Mondal, B.M. Bio-Chemistry and Nutritional health benefits of *Bellamya bengalensis* (Edible Gastropod): A Review. *Afr. J. Bio. Sci.* **2023**, *5*, 112–118.
13. Baghele, M.; Mishra, S.; Meyer-Rochow, V.B.; Jung, C.; Ghosh, S. A review of the nutritional potential of edible snails: A sustainable underutilized food resource. *Int. J. Nat. Prod. Res.* **2022**, *13*, 47930. [CrossRef]
14. Hayes, M.; Mora, L. Alternative Proteins as a Source of Bioactive Peptides: The Edible Snail and Generation of Hydrolysates Containing Peptides with Bioactive Potential for Use as Functional Foods. *Foods* **2021**, *10*, 276. [CrossRef]
15. Ortizo, R.G.G.; Sharma, V.; Tsai, M.L.; Wang, J.X.; Sun, P.P.; Nargotra, P.; Dong, C.D. Extraction of novel bioactive peptides from fish protein hydrolysates by enzymatic reactions. *Appl. Sci.* **2023**, *13*, 5768. [CrossRef]
16. Wu, Y.H.S.; Chen, Y.C. Trends and applications of food protein-origin hydrolysates and bioactive peptides. *J. Food Drug Anal.* **2022**, *30*, 172. [CrossRef]
17. Amigo, L.; Hernández-Ledesma, B. Current evidence on the bioavailability of food bioactive peptides. *Molecules* **2020**, *25*, 4479. [CrossRef]
18. Aguilar-Toalá, J.E.; Quintanar-Guerrero, D.; Liceaga, A.M.; Zambrano-Zaragoza, M.L. Encapsulation of bioactive peptides: A strategy to improve the stability, protect the nutraceutical bioactivity and support their food applications. *RSC Adv.* **2022**, *12*, 6449–6458. [CrossRef]
19. Peighambardoust, S.H.; Karami, Z.; Pateiro, M.; Lorenzo, J.M. A review on health-promoting, biological, and functional aspects of bioactive peptides in food applications. *Biomolecules* **2021**, *11*, 631. [CrossRef]
20. Mazloomi, N.; Safari, B.; Karaca, A.C.; Karimzadeh, L.; Moghadasi, S.; Ghanbari, M.; Assadpour, E.; Sarabandi, K.; Jafari, S.M. Loading bioactive peptides within different nanocarriers to enhance their functionality and bioavailability; in vitro and in vivo studies. *Adv. Colloid Interface Sci.* **2024**, *334*, 103318. [CrossRef]
21. Akbarbaglu, Z.; Peighambardoust, S.H.; Sarabandi, K.; Jafari, S.M. Spray drying encapsulation of bioactive compounds within protein-based carriers; different options and applications. *Food Chem.* **2021**, *359*, 129965. [CrossRef] [PubMed]
22. Culas, M.S.; Popovich, D.G.; Rashidinejad, A. Recent advances in encapsulation techniques for cinnamon bioactive compounds: A review on stability, effectiveness, and potential applications. *Food Biosci.* **2023**, *57*, 103470. [CrossRef]
23. Vaishnav, A.; Lal, J.; Mehta, N.K.; Mohanty, S.; Yadav, K.K.; Priyadarshini, M.B.; Singh, S.K. Unlocking the Potential of Fishery Waste: Exploring Diverse Applications of Fish Protein Hydrolysates in Food and Nonfood Sectors. *Environ. Sci. Pollut. Res.* **2025**, *22*, 1–45. [CrossRef] [PubMed]
24. Askari Vaselabadi, S.; Gharibzahedi, S.M.T.; Greiner, R.; Vale, J.M.; Ovenseri, A.C.; Rashidinejad, A.; Roohinejad, S. Advancements in Spray-Drying for the Microencapsulation of Fat-Soluble Vitamins: Stability, Bioavailability, and Applications. *J. Food Biochem.* **2025**, *2025*, 9974476. [CrossRef]

25. Baltrusch, K.L.; Torres, M.D.; Domínguez, H.; Flórez-Fernández, N. Spray-drying microencapsulation of tea extracts using green starch, alginate or carrageenan as carrier materials. *Int. J. Biol. Macromol.* **2022**, *203*, 417–429. [CrossRef]
26. Furuta, T.; Neoh, T.L. Microencapsulation of food bioactive components by spray drying: A review. *Dry. Technol.* **2021**, *39*, 1800–1831. [CrossRef]
27. Usov, A.I. Polysaccharides of the red algae. In *Advances in Carbohydrate Chemistry and Biochemistry*; Academic Press: Cambridge, MA, USA, 2011; Volume 65, pp. 115–217. [CrossRef]
28. Lomartire, S.; Gonçalves, A.M. Novel technologies for seaweed polysaccharides extraction and their use in food with therapeutically applications—A review. *Foods* **2022**, *11*, 2654. [CrossRef]
29. Inguglia, E.S.; Song, Z.; Kerry, J.P.; O'Sullivan, M.G.; Hamill, R.M. Addressing clean label trends in commercial meat processing: Strategies, challenges and insights from consumer perspectives. *Foods* **2023**, *12*, 2062. [CrossRef]
30. Jafari, S.; Jafari, S.M.; Ebrahimi, M.; Kijpatanasilp, I.; Assatarakul, K. A decade overview and prospect of spray drying encapsulation of bioactives from fruit products: Characterization, food application and in vitro gastrointestinal digestion. *Food Hydrocoll.* **2023**, *134*, 108068. [CrossRef]
31. Chakraborty, S. Carrageenan for encapsulation and immobilization of flavour, fragrance, probiotics, and enzymes: A review. *J. Carbohydr. Chem.* **2017**, *36*, 1–19. [CrossRef]
32. Mardani, M.; Siahtiri, S.; Besati, M.; Baghani, M.; Baniassadi, M.; Nejad, A.M. Microencapsulation of natural products using spray drying; an overview. *J. Microencapsul.* **2024**, *41*, 649–678. [CrossRef] [PubMed]
33. Vaishnav, A.; Mehta, N.K.; Hussain, S.A.; Acharya, P.C.; Biswal, S.; Nath, H.; Priyadarshini, M.B.; Thangarani, A.J.; Pal, P.; Singh, S.K.; et al. Bromelain excised hydrolysates with potent bioactivity from *Bellamya bengalensis* soft tissues: Process optimization and characterization. *J. Agric. Food Res.* **2025**, *17*, 101595. [CrossRef]
34. Hoyle, N.T.; Merritt, J.H. Quality of fish protein hydrolysates from herring (*Clupea harengus*). *J. Food Sci.* **1994**, *59*, 76–79. [CrossRef]
35. AOAC. *Official Method of Analysis of the AOAC International*, 20th ed.; Association of Official and Analytical Chemists International: Washington, VA, USA, 2016.
36. Lowry, O.H.; Rosebrough, N.J.; Farr, A.L.; Randall, R.J. Protein measurement with the Folin phenol reagent. *J. Biol. Chem.* **1951**, *193*, 265–275. [CrossRef]
37. Elavarasan, K.; Naveen Kumar, V.; Shamasundar, B.A. Antioxidant and functional properties of fish protein hydrolysates from fresh water carp (*Catla catla*) as influenced by the Nature of Enzyme. *J. Food Process. Preserv.* **2014**, *38*, 1207–1214. [CrossRef]
38. Mendanha, D.V.; Ortiz, S.E.M.; Favaro-Trindade, C.S.; Mauri, A.; Monterrey-Quintero, E.S.; Thomazini, M. Microencapsulation of casein hydrolysate by complex coacervation with SPI/pectin. *Food Res. Int.* **2009**, *42*, 1099–1104. [CrossRef]
39. Unnikrishnan, P.; Kizhakkethil, B.P.; George, J.C.; Aliyamveetil Abubacker, Z.; Ninan, G.; Nagarajarao, R.C. Antioxidant peptides from dark meat of yellowfin tuna (*Thunnus albacares*): Process optimization and characterization. *Waste Biomass Valorization* **2021**, *12*, 1845–1860. [CrossRef]
40. Zaitoun, B.J.; Palmer, N.; Amamcharla, J.K. Characterization of a Commercial Whey Protein Hydrolysate and Its Use as a Binding Agent in the Whey Protein Isolate Agglomeration Process. *Foods* **2022**, *11*, 1797. [CrossRef]
41. Yin, B.; Li, T.; Zhang, S.; Li, Z.; He, P. Sensitive analysis of 33 free amino acids in serum, milk, and muscle by ultrahigh-performance liquid chromatography-quadrupole-orbitrap high resolution mass spectrometry. *Food Anal. Methods* **2016**, *9*, 2814–2823. [CrossRef]
42. Klompong, V.; Benjakul, S.; Kantachote, D.; Shahidi, F. Antioxidative activity and functional properties of protein hydrolysate of yellow stripe trevally (*Selaroides leptolepis*) as influenced by the degree of hydrolysis and enzyme type. *Food Chem.* **2007**, *102*, 1317–1327. [CrossRef]
43. Gaurav, K.; Mehta, N.K.; Priyadarshini, M.B.; Pal, P.; Nirmal, N.P.; Sharma, S.; Vaishnav, A. Fat Uptake Reduction During Deep-Fat Frying of Fish Fillets Coated with Soy Protein Isolate Edible Coating. *J. Aquat. Food Prod. Technol.* **2024**, *33*, 573–588. [CrossRef]
44. Mohammadi, M.; Hamishehkar, H.; McClements, D.J.; Shahvalizadeh, R.; Barri, A. Encapsulation of Spirulina protein hydrolysates in liposomes: Impact on antioxidant activity and gastrointestinal behavior. *Food Chem.* **2023**, *400*, 133973. [CrossRef] [PubMed]
45. Pulido, R.; Bravo, L.; Saura-Calixto, F. Antioxidant activity of dietary polyphenols as determined by a modified ferric reducing/antioxidant power assay. *J. Agric. Food Chem.* **2000**, *48*, 3396–3402. [CrossRef] [PubMed]
46. You, L.; Zhao, M.; Liu, R.H.; Regenstein, J.M. Antioxidant and antiproliferative activities of loach (*Misgurnus anguillicaudatus*) peptides prepared by papain digestion. *J. Agric. Food Chem.* **2011**, *59*, 7948–7953. [CrossRef]
47. Soltanzadeh, M.; Peighambardoust, S.H.; Ghanbarzadeh, B.; Mohammadi, M.; Lorenzo, J.M. Chitosan nanoparticles as a promising nanomaterial for encapsulation of pomegranate (*Punica granatum* L.) peel extract as a natural source of antioxidants. *Nanomaterials* **2021**, *11*, 61439. [CrossRef]
48. Auwal, S.M.; Zarei, M.; Abdul-Hamid, A.; Saari, N. Optimization of bromelain-aided production of angiotensin I-converting enzyme inhibitory hydrolysates from stone fish using response surface methodology. *Mar. Drugs* **2017**, *15*, 104. [CrossRef]
49. Yi, D.; Lin, Q.; Johns, P.W. Estimation of degree of hydrolysis of protein hydrolysates by size exclusion chromatography. *Food Anal. Methods* **2021**, *14*, 805–813. [CrossRef]

50. Megrous, S.; Zhao, X.; Al-Dalali, S.; Yang, Z. Response surface methodology and optimization of hydrolysis conditions for the in vitro calcium-chelating and hypoglycemic activities of casein protein hydrolysates prepared using microbial proteases. *J. Food Meas. Charact.* **2024**, *18*, 3069–3084. [CrossRef]
51. da Rosa Zavareze, E.; Telles, A.C.; El Halal, S.L.M.; da Rocha, M.; Colussi, R.; de Assis, L.M.; Prentice-Hernández, C. Production and characterization of encapsulated antioxidative protein hydrolysates from Whitemouth croaker (*Micropogonias furnieri*) muscle and byproduct. *LWT-Food Sci. Technol.* **2014**, *59*, 841–848. [CrossRef]
52. Vioque, J.; Sánchez-Vioque, R.; Clemente, A.; Pedroche, J.; Millán, F. Partially hydrolysed rapeseed protein isolates with improved functional properties. *J. Am. Oil Chem. Soc.* **2000**, *77*, 447–450. [CrossRef]
53. Fenton, T.; Kanyuck, K.; Mills, T.; Pelan, E. Formulation and characterization of kappa-carrageenan gels with nonionic surfactant for melting-triggered controlled release. *Carbohydr. Polym. Technol. Appl.* **2021**, *2*, 100060. [CrossRef]
54. Udo, T.; Mummaleti, G.; Mohan, A.; Singh, R.K.; Kong, F. Current and emerging applications of carrageenan in the food industry. *Food Res. Int.* **2023**, *173*, 113369. [CrossRef]
55. Gu, L.; McClements, D.J.; Li, J.; Su, Y.; Yang, Y.; Li, J. Formulation of alginate/carrageenan microgels to encapsulate, protect and release immunoglobulins: Egg Yolk IgY. *Food Hydrocoll.* **2021**, *112*, 106349. [CrossRef]
56. Chuaychan, S.; Benjakul, S. Effect of maltodextrin on characteristics and antioxidative activity of spray-dried powder of gelatin and gelatin hydrolysate from scales of spotted golden goatfish. *J. Food Sci. Technol.* **2016**, *53*, 3583–3592. [CrossRef] [PubMed]
57. Cheng, Z.; Zhang, B.; Qiao, D.; Yan, X.; Zhao, S.; Jia, C.; Xu, Y. Addition of κ-carrageenan increases the strength and chewiness of gelatin-based composite gel. *Food Hydrocoll.* **2022**, *128*, 107565. [CrossRef]
58. Kokkuvayil Ramadas, B.; Rhim, J.W.; Roy, S. Recent progress of carrageenan-based composite films in active and intelligent food packaging applications. *Polymers* **2024**, *16*, 1001. [CrossRef]
59. Fang, Z.; Bhandari, B. Encapsulation of polyphenols—A review. *Trends Food Sci. Technol.* **2010**, *21*, 510–523. [CrossRef]
60. Kaveh, S.; Mahoonak, A.S.; Erfani-Moghadam, V.; Ghorbani, M.; Gholamhosseinpour, A.; Raeisi, M. Protein Hydrolysates of Niosome-Encapsulated Skipjack (*Katsuwonus pelamis*) Viscera: Antioxidant, Anticancer, and Release Properties. *Appl. Food Res.* **2025**, *5*, 100733. [CrossRef]
61. Sablania, V.; Bosco, S.J.D.; Rohilla, S.; Shah, M.A. Microencapsulation of *Murraya koenigii* L. leaf extract using spray drying. *J. Food Meas. Charact.* **2018**, *12*, 892–901. [CrossRef]
62. Permanadewi, I.; Kumoro, A.C.; Wardhani, D.H.; Aryanti, N. Effect of viscosity on iron encapsulation using alginate as a carrying agent in a controlled spray drying process. *Food Res.* **2022**, *6*, 56–67. [CrossRef]
63. Ren, D.; Xue, G.; Zheng, H.; Yang, W.; Cao, W.; Lin, H.; Gao, J.; Qin, X.; Zhang, C. Thermal stabilization effects of κ-Carrageenan on water-soluble protein extracted from *Pinctada martensii* meat. *J. Food Meas. Charact.* **2022**, *16*, 4985–4995. [CrossRef]
64. Annamalai, J.; Aliyamveetil Abubacker, Z.; Lakshmi, N.M.; Unnikrishnan, P. Microencapsulation of Fish Oil Using Fish Protein Hydrolysate, Maltodextrin, and Gum Arabic: Effect on Structural and Oxidative Stability. *J. Aquat. Food Prod. Technol.* **2020**, *29*, 293–306. [CrossRef]
65. Kurozawa, L.E.; Park, K.J.; Hubinger, M.D. Effect of Carrier Agents on the Physicochemical Properties of a Spray Dried Chicken Meat Protein Hydrolysate. *J. Food Eng.* **2009**, *94*, 326–333. [CrossRef]
66. Tonon, R.V.; Grosso, C.R.; Hubinger, M.D. Influence of emulsion composition and inlet air temperature on the microencapsulation of flaxseed oil by spray drying. *Food Res. Int.* **2011**, *44*, 282–289. [CrossRef]
67. Mavelil-Sam, R.; Ouseph, E.M.; Morreale, M.; Scaffaro, R.; Thomas, S. Recent developments and formulations for hydrophobic modification of carrageenan bionanocomposites. *Polymers* **2023**, *15*, 1650. [CrossRef]
68. Dong, Y.; Wei, Z.; Xue, C. Recent advances in carrageenan-based delivery systems for bioactive ingredients: A review. *Trends Food Sci. Technol.* **2021**, *112*, 348–361. [CrossRef]
69. Marín-Peñalver, D.; Alemán, A.; Montero, M.P.; Gómez-Guillén, M.C. Entrapment of natural compounds in spray-dried and heat-dried iota-carrageenan matrices as functional ingredients in surimi gels. *Food Funct.* **2021**, *12*, 2137–2147. [CrossRef]
70. Gómez-Mascaraque, L.G.; Llavata-Cabrero, B.; Martínez-Sanz, M.; Fabra, M.J.; López-Rubio, A. Self-assembled gelatin-ι-carrageenan encapsulation structures for intestinal-targeted release applications. *J. Colloid Interface Sci.* **2018**, *517*, 113–123. [CrossRef]
71. Cheng, C.; Chen, S.; Su, J.; Zhu, M.; Zhou, M.; Chen, T.; Han, Y. Recent advances in carrageenan-based films for food packaging applications. *Front. Nutr.* **2022**, *9*, 1004588. [CrossRef]
72. Sánchez, A.; Vázquez, A. Bioactive peptides: A review. *Food Qual. Saf.* **2017**, *1*, 29–46. [CrossRef]
73. Guidea, A.; Zăgrean-Tuza, C.; Moț, A.C.; Sârbu, C. Comprehensive evaluation of radical scavenging, reducing power, and chelating capacity of free proteinogenic amino acids using spectroscopic assays and multivariate exploratory techniques. *Spectrochim. Acta Part A Mol. Biomol. Spectrosc.* **2020**, *233*, 118158. [CrossRef] [PubMed]
74. Stipanuk, M.H. Metabolism of sulfur-containing amino acids: How the body copes with excess methionine, cysteine, and sulfide. *J. Nutr.* **2020**, *150*, 2494S–2505S. [CrossRef] [PubMed]

75. Zaky, A.A.; Simal-Gandara, J.; Eun, J.B.; Shim, J.H.; Abd El-Aty, A.M. Bioactivities, applications, safety, and health benefits of bioactive peptides from food and byproducts: A review. *Front. Nutr.* **2022**, *8*, 815640. [CrossRef]
76. Dimina, L.; Rémond, D.; Huneau, J.F.; Mariotti, F. Combining plant proteins to achieve amino acid profiles adapted to various nutritional objectives—An exploratory analysis using linear programming. *Front. Nutr.* **2022**, *8*, 809685. [CrossRef]
77. Feng, L.; Wu, Y.; Han, Y.; Yao, X.; Li, Q.; Liu, M.; Cao, Y. Structural characteristics, functional properties and nutritional value of walnut protein by limited enzymatic hydrolysis. *LWT* **2024**, *197*, 115923. [CrossRef]
78. Bielecka, M.; Cichosz, G.; Czeczot, H. Antioxidant, antimicrobial, and anticarcinogenic activities of bovine milk proteins and their hydrolysates—A review. *Int. Dairy J.* **2022**, *127*, 105208. [CrossRef]
79. McClements, D.J. Nanoscale nutrient delivery systems for food applications: Improving bioactive dispersibility, stability, and bioavailability. *J. Food Sci.* **2015**, *80*, N1602–N1611. [CrossRef]
80. Yeerong, K.; Chantawannakul, P.; Anuchapreeda, S.; Juntrapirom, S.; Kanjanakawinkul, W.; Müllertz, A.; Chaiyana, W. Chitosan Alginate Nanoparticles of Protein Hydrolysate from *Acheta domesticus* with Enhanced Stability for Skin Delivery. *Pharmaceutics* **2024**, *16*, 724. [CrossRef]
81. Krempel, M.; Griffin, K.; Khouryieh, H. 13-Hydrocolloids as Emulsifiers and Stabilizers in Beverage Preservation. In *Preservatives and Preservation Approaches in Beverages*; Academic Press: Cambridge, MA, USA, 2019; pp. 427–465. [CrossRef]
82. Song, X.; Chen, Y.; Sun, H.; Liu, X.; Leng, X. Physicochemical stability and functional properties of selenium nanoparticles stabilized by chitosan, carrageenan, and gum Arabic. *Carbohydr. Polym.* **2021**, *255*, 117379. [CrossRef]
83. Li, X.; Chen, X.; Cheng, H. Impact of κ-Carrageenan on the Cold-Set Pea Protein Isolate Emulsion-Filled Gels: Mechanical Property, Microstructure, and In Vitro Digestive Behavior. *Foods* **2024**, *13*, 483. [CrossRef] [PubMed]
84. Li, J.; Zhu, F. Whey protein hydrolysates and infant formulas: Effects on physicochemical and biological properties. *Compr. Rev. Food Sci. Food Saf.* **2024**, *23*, e13337. [CrossRef] [PubMed]
85. Masarudin, M.J.; Cutts, S.M.; Evison, B.J.; Phillips, D.R.; Pigram, P.J. Factors determining the stability, size distribution, and cellular accumulation of small, monodisperse chitosan nanoparticles as candidate vectors for anticancer drug delivery: Application to the passive encapsulation of [14C]-doxorubicin. *Nanotechnol. Sci. Appl.* **2015**, *67*, 67–80. [CrossRef] [PubMed]
86. Danaei, M.R.M.M.; Dehghankhold, M.; Ataei, S.; Hasanzadeh Davarani, F.; Javanmard, R.; Dokhani, A.; Khorasani, S.; Mozafari, M.R. Impact of particle size and polydispersity index on the clinical applications of lipidic nanocarrier systems. *Pharmaceutics* **2018**, *10*, 57. [CrossRef] [PubMed]
87. Zabot, G.L.; Schaefer Rodrigues, F.; Polano Ody, L.; Vinícius Tres, M.; Herrera, E.; Palacin, H.; Córdova-Ramos, J.S.; Best, I.; Olivera-Montenegro, L. Encapsulation of bioactive compounds for food and agricultural applications. *Polymers* **2022**, *14*, 4194. [CrossRef]
88. Wang, Y.H.; Wang, J.M.; Wan, Z.L.; Yang, X.Q.; Chen, X.W. Corn protein hydrolysate as a new structural modifier for soybean protein isolate based O/W emulsions. *LWT* **2020**, *118*, 108763. [CrossRef]
89. Soukoulis, C.; Bohn, T. A comprehensive overview on the micro and nanotechnological encapsulation advances for enhancing the chemical stability and bioavailability of carotenoids. *Crit. Rev. Food Sci. Nutr.* **2018**, *58*, 1–36. [CrossRef]
90. Đorđević, V.; Balanč, B.; Belščak-Cvitanović, A.; Lević, S.; Trifković, K.; Kalušević, A.; Kostić, I.; Komes, D.; Bugarski, B.; Nedović, V. Trends in encapsulation technologies for delivery of food bioactive compounds. *Food Eng. Rev.* **2015**, *7*, 452–490. [CrossRef]
91. Hosseini, S.F.; Ramezanzade, L.; Nikkhah, M. Nano-Liposomal Entrapment of Bioactive Peptidic Fraction from Fish Gelatin Hydrolysate. *Int. J. Biol. Macromol.* **2017**, *105*, 1455–1463. [CrossRef] [PubMed]
92. Midekessa, G.; Godakumara, K.; Ord, J.; Viil, J.; Lättekivi, F.; Dissanayake, K.; Kopanchuk, S.; Rinken, A.; Andronowska, A.; Bhattacharjee, S.; et al. Zeta potential of extracellular vesicles: Toward understanding the attributes that determine colloidal stability. *ACS Omega* **2020**, *5*, 16701–16710. [CrossRef]
93. Liu, J.; Zhan, X.; Wan, J.; Wang, Y.; Wang, C. Review for carrageenan-based pharmaceutical biomaterials: Favourable physical features versus adverse biological effects. *Carbohydr. Polym.* **2015**, *121*, 27–36. [CrossRef] [PubMed]
94. Feng, Y.; Kilker, S.R.; Lee, Y. Surface charge (zeta-potential) of nanoencapsulated food ingredients. In *Characterization of Nanoencapsulated Food Ingredients*; Academic Press: Cambridge, MA, USA, 2020; pp. 213–241. [CrossRef]
95. Yousefi, N.; Abbasi, S. Food proteins: Solubility & thermal stability improvement techniques. *Food Chem. Adv.* **2022**, *1*, 100090. [CrossRef]
96. Alu'datt, M.H.; Alrosan, M.; Gammoh, S.; Tranchant, C.C.; Alhamad, M.N.; Rababah, T.; Zghoul, R.; Alzoubi, H.; Ghatasheh, S.; Ghozlan, K.; et al. Encapsulation-based technologies for bioactive compounds and their application in the food industry: A roadmap for food-derived functional and health-promoting ingredients. *Food Biosci.* **2022**, *50*, 101971. [CrossRef]
97. Rivera-Jiménez, J.; Berraquero-García, C.; Pérez-Gálvez, R.; García-Moreno, P.J.; Espejo-Carpio, F.J.; Guadix, A.; Guadix, E.M. Peptides and protein hydrolysates exhibiting anti-inflammatory activity: Sources, structural features, and modulation mechanisms. *Food Funct.* **2022**, *13*, 12510–12540. [CrossRef]
98. Berraquero-García, C.; Pérez-Gálvez, R.; Espejo-Carpio, F.J.; Guadix, A.; Guadix, E.M.; García-Moreno, P.J. Encapsulation of bioactive peptides by spray-drying and electrospraying. *Foods* **2023**, *12*, 2005. [CrossRef]

99. Gharanjig, H.; Gharanjig, K.; Hosseinnezhad, M.; Jafari, S.M. Differential scanning calorimetry (DSC) of nanoencapsulated food ingredients. In *Characterization of Nanoencapsulated Food Ingredients*; Academic Press: Cambridge, MA, USA, 2020; pp. 295–346. [CrossRef]
100. Sooram, B.; Gupta, N.; Chethireddy, V.R.; Tripathi, T.; Saudagar, P. Applications of Differential Scanning Calorimetry in Studying Folding and Stability of Proteins. In *Protein Folding Dynamics and Stability: Experimental and Computational Methods*; Springer Nature: Singapore, 2023; pp. 37–60. [CrossRef]
101. TIP, T. Interpreting DSC Curves Part 1: Dynamic Measurements. Available online: https://www.eng.uc.edu/~beaucag/Classes/Characterization/DSCParts/Artifacts%20in%20DSC%20Usercom_11.pdf (accessed on 20 February 2025).
102. Zhou, H.X.; Pang, X. Electrostatic interactions in protein structure, folding, binding, and condensation. *Chem. Rev.* **2018**, *118*, 1691–1741. [CrossRef]
103. Yang, X.; Lv, Z.; Han, C.; Zhang, J.; Duan, Y.; Guo, Q. Stability and encapsulation properties of daidzein in zein/carrageenan/sodium alginate nanoparticles with ultrasound treatment. *Int. J. Biol. Macromol.* **2024**, *262*, 130070. [CrossRef]
104. Yathisha, U.G.; Vaidya, S.; Sheshappa, M.B. Functional properties of protein hydrolyzate from ribbon fish (*Leptura canthus savala*) as-prepared by enzymatic hydrolysis. *Int. J. Food Prop.* **2022**, *25*, 187–203. [CrossRef]
105. Patil, U.; Gulzar, S.; Ma, L.; Zhang, B.; Benjakul, S. Pickering emulsion stabilized by fish myofibrillar proteins modified with tannic acid, as influenced by different drying methods. *Foods* **2023**, *12*, 1556. [CrossRef]
106. Sun, W.; Zhou, F.; Zhao, M.; Yang, B.; Cui, C. Physicochemical changes of myofibrillar proteins during processing of Cantonese sausage in relation to their aggregation behavior and in vitro digestibility. *Food Chem.* **2011**, *129*, 472–478. [CrossRef]
107. Wachirattanapongmetee, K.; Katekaew, S.; Weerapreeyakul, N.; Thawornchinsombut, S. Differentiation of protein types extracted from tilapia byproducts by FTIR spectroscopy combined with chemometric analysis and their antioxidant protein hydrolysates. *Food Chem.* **2024**, *437*, 137862. [CrossRef] [PubMed]
108. Muyonga, J.H.; Cole, C.G.B.; Duodu, K.G. Fourier transform infrared (FTIR) spectroscopic study of acid-soluble collagen and gelatin from skins and bones of young and adult Nile perch (*Lates niloticus*). *Food Chem.* **2004**, *86*, 325–332. [CrossRef]
109. Webber, V.; de Carvalho, S.M.; Ogliari, P.J.; Hayashi, L.; Barreto, P.L.M. Optimization of the extraction of carrageenan from *Kappaphycus alvarezii* using response surface methodology. *Food Sci. Technol.* **2012**, *32*, 812–818. [CrossRef]
110. Sadat, A.; Joye, I.J. Peak fitting applied to Fourier transform infrared and Raman spectroscopic analysis of proteins. *Appl. Sci.* **2020**, *10*, 5918. [CrossRef]
111. Sarabandi, K.; Gharehbeglou, P.; Jafari, S.M. Spray-drying encapsulation of protein hydrolysates and bioactive peptides: Opportunities and challenges. *Dry. Technol.* **2020**, *38*, 577–595. [CrossRef]
112. Vicente, J.; Pinto, J.; Menezes, J.; Gaspar, F. Fundamental analysis of particle formation in spray drying. *Powder Technol.* **2013**, *247*, 1–7. [CrossRef]
113. Gomes, M.H.G.; Kurozawa, L.E. Surface composition and microstructure of spray-dried microparticles: A review about their effect on functional properties. *Dry. Technol.* **2024**, *42*, 1665–1682. [CrossRef]
114. Bakry, A.M.; Huang, J.; Zhai, Y.; Huang, Q. Myofibrillar protein with κ- or λ-carrageenans as novel shell materials for microencapsulation of tuna oil through complex coacervation. *Food Hydrocoll.* **2019**, *96*, 43–53. [CrossRef]
115. Garzón, A.G.; Cian, R.E.; Drago, S.R. Effects of agar-carrageenan wall materials and core-to-wall material ratio on physicochemical properties and in vitro bioaccessibility of microencapsulated bioactive peptides. *Food Hydrocoll.* **2023**, *139*, 108570. [CrossRef]
116. McClements, D.J. Recent developments in encapsulation and release of functional food ingredients: Delivery by design. *Curr. Opin. Food Sci.* **2018**, *23*, 80–84. [CrossRef]
117. Klojdová, I.; Milota, T.; Smetanová, J.; Stathopoulos, C. Encapsulation: A strategy to deliver therapeutics and bioactive compounds? *Pharmaceutics* **2023**, *16*, 362. [CrossRef] [PubMed]
118. Slizyte, R.; Rommi, K.; Mozuraityte, R.; Eck, P.; Five, K.; Rustad, T. Bioactivities of fish protein hydrolysates from defatted salmon backbones. *Biotechnol. Rep.* **2016**, *11*, 99–109. [CrossRef] [PubMed]
119. Klomklao, S.; Benjakul, S. Protein hydrolysates prepared from the viscera of skipjack tuna (*Katsuwonus pelamis*): Antioxidative activity and functional properties. *Turk. J. Fish. Aquat. Sci.* **2018**, *18*, 69–79. [CrossRef] [PubMed]
120. Yarnpakdee, S.; Benjakul, S.; Kristinsson, H.G.; Kishimura, H. Antioxidant and sensory properties of protein hydrolysate derived from Nile tilapia (*Oreochromis niloticus*) by one-and two-step hydrolysis. *J. Food Sci. Technol.* **2015**, *52*, 3336–3349. [CrossRef]
121. Xue, J.; Luo, Y. Protein-polysaccharide nanocomplexes as nanocarriers for delivery of curcumin: A comprehensive review on preparation methods and encapsulation mechanisms. *J. Future Foods* **2023**, *3*, 99–114. [CrossRef]
122. Pourashouri, P.; Shabanpour, B.; Heydari, S.; Raeisi, S. Encapsulation of fish oil by carrageenan and gum tragacanth as wall materials and its application to the enrichment of chicken nuggets. *LWT* **2021**, *137*, 110334. [CrossRef]
123. Chalamaiah, M.; Jyothirmayi, T.; Diwan, P.V.; Kumar, B.D. Antioxidant activity and functional properties of enzymatic protein hydrolysates from common carp (*Cyprinus carpio*) roe (egg). *J. Food Sci. Technol.* **2015**, *52*, 5817–5825. [CrossRef]

124. Tadesse, S.A.; Emire, S.A.; Barea, P.; Illera, A.E.; Melgosa, R.; Beltrán, S.; Sanz, M.T. Valorization of low-valued ray-finned fish (*Labeo barbus nedgia*) by enzymatic hydrolysis to obtain fish-discarded protein hydrolysates as functional foods. *Food Bioprod. Process.* **2023**, *141*, 167–184. [CrossRef]
125. Pimentel, F.B.; Machado, M.; Cermeño, M.; Kleekayai, T.; Machado, S.; Rego, A.M.; FitzGerald, R.J. Enzyme-assisted release of antioxidant peptides from *Porphyra dioica* conchocelis. *Antioxidants* **2021**, *10*, 249. [CrossRef]
126. Slizyte, R.; Mozuraityte, R.; Remman, T.; Rustad, T. Two-stage processing of salmon backbones to obtain high-quality oil and proteins. *Int. J. Food Sci. Technol.* **2018**, *53*, 2378–2385. [CrossRef]
127. Nemati, M.; Shahosseini, S.R.; Ariaii, P. Review of fish protein hydrolysates: Production methods, antioxidant and antimicrobial activity and nanoencapsulation. *Food Sci. Biotechnol.* **2024**, *33*, 1789–1803. [CrossRef] [PubMed]
128. Valko, M.; Leibfritz, D.; Moncol, J.; Cronin, M.T.; Mazur, M.; Telser, J. Free radicals and antioxidants in normal physiological functions and human disease. *Int. J. Biochem. Cell Biol.* **2007**, *39*, 44–84. [CrossRef] [PubMed]
129. Afzal, S.; Abdul Manap, A.S.; Attiq, A.; Albokhadaim, I.; Kandeel, M.; Alhojaily, S.M. From imbalance to impairment: The central role of reactive oxygen species in oxidative stress-induced disorders and therapeutic exploration. *Front. Pharmacol.* **2023**, *14*, 1269581. [CrossRef] [PubMed]
130. Shen, Q.; Ou, A.; Liu, S.; Elango, J.; Wang, S.; Henriques da Silva, T.; Bao, B. Effects of ion concentrations on the hydroxyl radical scavenging rate and reducing power of fish collagen peptides. *J. Food Biochem.* **2019**, *43*, e12789. [CrossRef]
131. Xu, C.; Cai, C.; Liu, T.; Kang, J.; Li, S.; Zhang, J.; Wang, H. Enhancing the antioxidant activity of fish scale collagen hydrolysates through Plastein reaction. *Food Bioprocess Technol.* **2024**, *17*, 3693–3703. [CrossRef]
132. Halliwell, B.; Gutteridge, J.M. *Free Radicals in Biology and Medicine*; Oxford University Press: New York, NY, USA, 2015.
133. Sarmadi, B.H.; Ismail, A. Antioxidative peptides from food proteins: A review. *Peptides* **2010**, *31*, 1949–1956. [CrossRef]
134. Cruz-Casas, D.E.; Ramos-González, R.; Prado-Barragán, L.A.; Aguilar, C.N.; Rodríguez-Herrera, R.; Iliná, A.; Esparza-González, S.C.; Flores-Gallegos, A.C. Antihypertensive amaranth protein hydrolysates encapsulation in alginate/pectin beads: Influence on bioactive properties upon in vitro digestion. *Polysaccharides* **2024**, *5*, 450–462. [CrossRef]
135. Rezagholizade-Shirvan, A.; Soltani, M.; Shokri, S.; Radfar, R.; Arab, M.; Shamloo, E. Bioactive compound encapsulation: Characteristics, applications in food systems, and implications for human health. *Food Chem. X* **2024**, *24*, 101953. [CrossRef]
136. Ilyasov, I.R.; Beloborodov, V.L.; Selivanova, I.A.; Terekhov, R.P. ABTS/PP decolorization assay of antioxidant capacity reaction pathways. *Int. J. Mol. Sci.* **2020**, *21*, 1131. [CrossRef]
137. Jemil, I.; Jridi, M.; Nasri, R.; Ktari, N.; Salem, R.B.S.B.; Mehiri, M.; Hajji, M.; Nasri, M. Functional, antioxidant, and antibacterial properties of protein hydrolysates prepared from fish meat fermented by *Bacillus subtilis* A26. *Process Biochem.* **2014**, *49*, 963–972. [CrossRef]
138. Rivero-Pino, F.; Espejo-Carpio, F.J.; Guadix, E.M. Evaluation of the Bioactive Potential of Foods Fortified with Fish Protein Hydrolysates. *Food Res. Int.* **2020**, *137*, 109572. [CrossRef]
139. Xia, W.; Gao, Y.; Fang, X.; Jin, L.; Liu, R.; Wang, L.S.; Deng, Y.; Gao, J.; Yang, H.; Wu, W.; et al. Simulated Gastrointestinal Digestion of Walnut Protein Yields Anti-Inflammatory Peptides. *Food Chem.* **2024**, *445*, 138646. [CrossRef]

colloids and interfaces

MDPI

Article

Ethanolic Cashew Leaf Extract Encapsulated in Tripolyphosphate–Chitosan Complexes: Characterization, Antimicrobial, and Antioxidant Activities

Pitima Sinlapapanya [1], Jirayu Buatong [1], Suriya Palamae [1], Rasool Abdul Nazeer [2], Bin Zhang [3], Thummanoon Prodpran [4] and Soottawat Benjakul [1,5,*]

1 International Center of Excellence in Seafood Science and Innovation, Faculty of Agro-Industry, Prince of Songkla University, Hat Yai, Songkhla 90110, Thailand; pitimasinlapapanya@gmail.com (P.S.); jirayu.b@psu.ac.th (J.B.); suriya.pal@psu.ac.th (S.P.)
2 Biopharmaceuticals Lab, Department of Biotechnology, School of Bioengineering, SRM Institute of Science and Technology, Kattankulathur 603203, Tamilnadu, India; nazeerr@srmist.edu.in
3 Key Laboratory of Health Risk Factors for Seafood of Zhejiang Province, College of Food Science and Pharmacy, Zhejiang Ocean University, Zhoushan 316022, China; zhangbin_ouc@163.com
4 Department of Material Product Technology, Faculty of Agro-Industry, Prince of Songkla University, Hat Yai, Songkhla 90110, Thailand; thummanoon.p@psu.ac.th
5 Department of Food and Nutrition at Kyung Hee University, Seoul 02447, Republic of Korea
* Correspondence: soottawat.b@psu.ac.th

Abstract: Ethanolic cashew leaf extract (ECL-E) is rich in phenolic compounds and shows remarkable antioxidative and antimicrobial activities. Encapsulation could stabilize ECL-E as the core. Tripolyphosphate (TPP)–chitosan (CS) nanoparticles were used to load ECL-E, and the resulting nanoparticles were characterized. The nanoparticles loaded with ECL-E at different levels showed differences in encapsulation efficiency (47.62–89.47%), mean particle diameters (47.30–314.60 nm), positive zeta potentials (40.37–44.24 mV), and polydispersity index values (0.20–0.56). According to scanning electron micrographs, the nanoparticles had a spherical or ellipsoidal shape, and a slight agglomeration was observed. The appropriate ratio of CS/ECL-E was 1:3, in which an EE of 89.47%, a particle size of 256.05 ± 7.70 nm, a zeta potential of 40.37 ± 0.66 mV, and a PDI of 0.22 ± 0.05 were obtained. The nanoparticles also exhibited high antioxidant activities, as assayed by DPPH and ABTS radical scavenging activities, ferric reducing ability power (FRAP), and oxygen radical absorbance capacity (ORAC). Low minimum inhibitory concentration and minimum bactericidal concentration were observed against *Pseudomonas aeruginosa* (9.38, 75.00 mg/mL) and *Shewanella putrefaciens* (4.69, 75.00 mg/mL). In addition, ECL-E loaded in nanoparticles could maintain its bioactivities under various light intensities (1000–4000 Lux) for 48 h. Some interactions among TPP, CS, and ECL-E took place, as confirmed by FTIR analysis. These nanoparticles had the increased storage stability and could be used for inactivating spoilage bacteria and retarding lipid oxidation in foods.

Keywords: cashew leaf extract; encapsulation; chitosan; antimicrobial; antioxidant; nanoparticle

1. Introduction

Phenolic compounds are phytochemicals abundant in several plants, especially in cashew leaf extracts, that exhibit a wide range of biological activities [1,2]. Cashew leaf extract shows antioxidant and antimicrobial activities and can be used as a natural safe food additive [2,3]. However, some polyphenols are unstable and lose their bioavailability, which can be overcome through encapsulation [4,5].

Some bioactive substances and nutraceuticals need protection from external factors such as light, pH, temperature, oxygen, and enzymes. It is practicable to use micro- or nano-encapsulation to protect these substances [6]. Encapsulation in the form of microcapsules is an efficient method for safeguarding bioactive substances [7]. Encapsulated nutraceuticals or bioactive compounds can be added to food matrices to create functional meals that are highly absorbed by the digestive system [8]. Bioactive compounds are loaded inside microcapsules during the encapsulation process, and they are then released at a specific time [9]. To encapsulate bioactive compounds and nutraceuticals, nano-encapsulation has drawn attention to creating stable nutrients for enriched foods [10]. Biodegradable polymers such as polylactic acid (PLA), poly(lactic-co-glycolic acid) (PLGA), and poly(glycerol adipate) (PGA) have been used to encapsulate bioactive compounds. However, there are still limitations related to low loading efficiency. Some of the produced compounds have side effects (toxicity), and these polymers have complicated synthesis steps [11]. Since several polysaccharides are highly modifiable, soluble, and capable of binding via their functional groups, they are appropriate for the encapsulation process as wall materials [7,10].

Because of its low toxicity, high biodegradability, and antibacterial qualities, chitosan, a linear polysaccharide produced by partially deacetylating chitin, has attracted special attention for a wide range of applications [12–15]. Furthermore, because its chain contains several functional groups, chitosan is easily modifiable to achieve appropriate physicochemical properties [12]. Chitosan is a widely used biocompatible polysaccharide for the development of pharmaceutical formulations such as gels, films, tissue engineering scaffolds, and nanoparticles. It also has antimicrobial properties [16].

The cationic character of chitosan is modified by a polyanion, like sodium tripolyphosphate (TPP), which causes ionotropic gelation, and the subsequent production of nanoparticles is achieved [17,18]. Bioactive substances have been encapsulated in chitosan nanoparticles and employed as delivery nanocarriers [19,20]. It has been demonstrated that chitosan nanoparticles have antibacterial properties, thermal stability, antioxidative activity, aqueous medium solubility, and increased release of bioactive chemicals [21,22]. As such, TPP-CS nanoparticles have potential applications as nanocarriers in food.

Cashew is an abundant plant in tropical areas, especially in the southern part of Thailand. Its nut has a high price and has become popular for consumers. As a consequence, the leaves are also available for the extraction of active compounds. Recently, cashew leaf extract has been reported to contain a wide range of polyphenolic compounds such as amentoflavone, quercetin, kaempferol, and catechin [2]. In addition, according to the previous report by Sinlapapanya et al. [3], it was found that phenolic compounds easily lose their activities when stored because of environmental factors such as light and heat. To maintain the bioactivities of the extract, coacervation could be a promising means to stabilize an extract exposed to the environment. The aim of the present study was to encapsulate ECL-E in TPP-CS nanoparticles. Nanoparticles were then characterized, and the antioxidant and antimicrobial activities of ECL-E loaded in nanoparticles were determined.

2. Materials and Methods

2.1. Chemicals

Ethanol was supplied by RCI Labscan Limited (Bangkok, Thailand). Chitosan (low molecular weight, 50–190 kDa) and sodium tripolyphosphate (TPP) were acquired from Sigma Aldrich, Inc. (St. Louis, MO, USA). Mueller–Hinton broth (MHB) and tryptic soy broth were obtained from Himedia (Mumbai, India).

2.2. Preparation of Ethanolic Cashew Leaf Extract (ECL-E) and the Encapsulation Process

2.2.1. Preparation of ECL-E

Dried cashew (*Anacardium occidentale* L.) leaves were powderized before extraction using 80% ethanol following the method explained by Sinlapapanya et al. [2]. Leaf powder (10 g) was mixed with 80% ethanol (200 mL) before ultrasonication at 70% amplitude for 40 min under pulsed mode (5s:5s) (Sonics, Model VC750, Sonica & Materials, Inc., Newtown, CT, USA). The temperature of the mixture during extraction was maintained below 35 °C. After that, the supernatant was collected and centrifuged (5000× *g*, 30 min, 4 °C) (Beckman Coulter, Avanti J-E Centrifuge, Fullerton, CA, USA). An Eyela rotary evaporator (Tokyo Rikakikai, Co. Ltd., Tokyo, Japan) was employed to remove ethanol, and nitrogen flushing was applied to eliminate the remaining ethanol.

2.2.2. Dechlorophyllization of ECL-E

Chlorophylls in ECL-E were eliminated by the sedimentation process, as explained by Sinlapapanya et al. [3]. The extract and distilled water were mixed at 1:1 (v/v) ratio and properly mixed. After being chilled for 24 h, centrifugation (10,000× g, 30 min) was performed at 4 °C. The supernatant was freeze-dried and the final powder, known as "ethanolic cashew leaf extract" or "ECL-E", was packed in a zip-lock bag and kept at −20 °C.

2.2.3. TPP–Chitosan (TPP-CS) Encapsulation Process

Encapsulation via TPP-CS complexation was performed [7]. CS powder was dissolved in acetic acid solution (1% v/v) to gain the concentration of 1% (w/v). The mixture was stirred overnight, and centrifugation (9000× g, 10 min, 4 °C) was employed to remove the debris. ECL-E was mixed with CS solution (20 mL) to obtain varying CS/ECL-E ratios (1:1; 1:2; and 1:3) with the aid of homogenizer (IKA Labortechnik homogenizer, Selangor, Malaysia) at 13,000 rpm (Table 1). After pH adjustment to 4.2 with 2 M NaOH solution, TPP solution (0.4% w/v, 8 mL) was added dropwise with rigorous stirring to induce ionic gelation. Centrifugation (9000× g, 30 min, 4 °C) was used to separate TPP-CS gels from non-gelling substances, and the gels were dispersed in 100 mL of DI water containing 1% (w/v) trehalose. Then, the nanoparticles were prepared by dissociation of the flocculated pellet by ultrasonication (70% amplitude, 10 min) in pulse mode (5s:5s). TPP-CS complexes and TPP-CS complexes loaded with ECL-E were frozen (−40 °C) and lyophilized at −50 °C for 72 h by a freeze dryer (CoolSafe 55, ScanLaf A/S, Lynge, Denmark). Nanoparticles with CS/ECL-E ratios of 1:1, 1:2, and 1:3 were named as ECL-E-N-E1, ECL-E-N-E2, and ECL-E-N-E3, respectively. The nanoparticles were then analyzed.

Table 1. Encapsulation condition of ECL-E by the TPP-CS process.

Treatment	Ratio	EE (%)	PS (nm)	ZP (mV)	PDI	Color		
						L*	a*	b*
ECL-E	-	-	-	-	-	59.34 ± 0.07^{e}	1.33 ± 0.03^{a}	10.16 ± 0.04^{a}
TPP-CS	CS	43.01 ± 8.89^{b}	47.30 ± 4.60^{d}	43.96 ± 0.96^{a}	0.56 ± 0.06^{a}	63.02 ± 0.36^{d}	-0.54 ± 0.06^{d}	-0.65 ± 0.01^{d}
ECL-E-N-E1	CS(1):ECL-E(1)	47.62 ± 5.71^{b}	314.60 ± 20.65^{a}	44.24 ± 0.92^{a}	0.30 ± 0.03^{b}	64.71 ± 0.03^{c}	0.73 ± 0.05^{c}	10.07 ± 0.01^{b}
ECL-E-N-E2	CS(1):ECL-E(2)	86.72 ± 2.28^{a}	273.60 ± 3.70^{b}	41.13 ± 1.01^{b}	0.20 ± 0.02^{c}	66.91 ± 0.03^{b}	1.26 ± 0.02^{b}	9.73 ± 0.09^{c}
ECL-E-N-E3	CS(1):ECL-E(3)	89.47 ± 1.78^{a}	256.05 ± 7.70^{c}	40.37 ± 0.66^{b}	0.22 ± 0.05^{c}	69.55 ± 0.04^{a}	1.22 ± 0.05^{b}	10.10 ± 0.01^{a}

CS: chitosan, TPP: sodium tripolyphosphate, ECL-E: ethanolic cashew leaf extract, EE: encapsulation efficiency, PS: particle size, ZP: zeta potential, PDI: polydispersity index. Different lowercase superscripts within the same column indicate significant differences ($p < 0.05$). * Mean value ($n = 3$).

2.3. *Analyses*

2.3.1. Encapsulation Efficiency (EE)

The suspension of ECL-E-N was stirred gently for 30 min and centrifuged at 10,000× g for 15 min and 4 °C. The initial TPC of the dissolved ECL-E-N (M2) was noted.

$$EE\ (\%) = [1 - (M1/M2)] \times 100$$

where M1 is the TPC outside the dialysis bag (in 200 mL of dialysis water) and M2 is the total TPC loaded.

2.3.2. Particle Size (PS), Polydispersity Index (PDI) and Zeta Potential Measurement

The dynamic light scattering technique was used to determine the PS, PDI, and zeta potential of different samples by the ZetaPlus zeta potential analyzer (Brookhaven Instruments Corporation, Holtsville, NY, USA) according to Tagrida et al. [23]. Ten milliliters of each sample suspension (10 mg/mL) was appropriately diluted in DI water and measured at 25 °C with a medium refractive index of 1.330. The particle size and PDI were determined at an angle of 90 °C.

2.3.3. Color

A colorimeter (ColorFlex, Hunter Lab Reston, VA, USA) was employed to examine the lightness (L*), redness/greenness (a*), and yellowness/blueness (b*) of different nanoparticles.

2.3.4. In Vitro Releasing Efficiency (RE)

A dialysis bag (2.5 × 10 cm, MW cut-off: 3500 Da) was used to examine the in vitro RE of different ECL-E-Ns compared with ECL-E [24]. One mL of 1% ECL-E (control) and various ECL-E-Ns (1 mg/mL, 100 mg/mL) were transferred into each dialysis bag, and the bags were clamped before dialysis in 100 mL of DI water by continuous stirring (100 rpm) for 72 h at 4 °C. Three milliliters of dialysis water were taken every 12 h up to 72 h for TPC analysis. RE was determined using the equation shown below.

$$\text{RE (\%)} = [1 - (\text{TPC}_{\text{release}}/\text{TPC}_{\text{total}})] \times 100$$

where RE denotes the release efficiency (%), $\text{TPC}_{\text{release}}$ (mg/mL) represents the TPC released after being dialyzed at the designated time (every 12 h), and $\text{TPC}_{\text{total}}$ (mg/mL) is the total TPC in ECL-E-Ns, obtained from the known amount of TPC in ECL-E used for preparation of nanoparticles.

For TPC determination, the Folin–Ciocalteu's reagent (FCR) method was adopted as described by Benjakul et al. [25]. TPC was expressed as mg gallic acid equivalent (GAE)/g dry solid.

2.3.5. Antioxidant Activities

ECL-E-Ns with different loading conditions (CS/ECL-E ratios) were tested for antioxidant activities in comparison with ECL-E (unencapsulated) [3].

ABTS Radical Scavenging Activity

A sample (150 µL) was mixed with ABTS solution (2850 µL) and incubated for 1 h at room temperature in the dark. Absorbance 734 nm was read. Blank was prepared using distilled water instead of a sample.

DPPH Radical Scavenging Activity

A sample (0.3 mL) was mixed with 2.7 mL of 0.15 mM DPPH in methanol, shaken, and incubated for 60 min at 25 °C in the dark. DPPH reduction was monitored by the decrease in OD at 517 nm.

Ferric Reducing Ability Power

A sample solution (150 µL) was mixed with 2.85 mL of working FRAP reagent, and the mixture was incubated in the dark at room temperature for 30 min. The absorbance at 593 nm was read.

Metal Chelating Activity

A sample (1 mL) was mixed with distilled water (3.7 mL). Subsequently, 2 mM $FeCl_2$ (0.1 mL) and 5 mM ferrozine (0.2 mL) were added. The absorbance at 562 nm was read after 20 min. For the control, 1 mL of distilled water was used to replace the extract.

Oxygen Radical Absorbance Capacity

The oxygen radical absorbance capacity (ORAC) assay was performed according to Sae-Leaw et al. [26].

Antioxidant activities were expressed as µmol Trolox equivalents (TE)/g dry solid, except metal chelating activity, which was reported as µmol EDTA equivalent (EE)/g dry solid.

The nanoparticle with the optimal CS/ECL-E ratio and the highest antioxidant activity was chosen for further analysis.

2.3.6. Antimicrobial Activity

Pseudomonas aeruginosa ATCC 27853, obtained from The American Type Culture Collection, and *Shewanella putrefaciens* JCM 20190, from the Japan Collection of Microorganisms, were used. Both bacteria are major spoilage bacteria commonly found in seafood. All bacteria were checked for purity before being used each time by re-streaking from the stock on TSA agar and selecting only a single colony for enrichment.

Minimum Inhibitory Concentration (MIC) and Minimum Bactericidal Concentration (MBC)

A Mueller–Hinton broth microdilution was adopted to evaluate the MIC and MBC of the ECL-E-Ns in a 96-well microplate [2]. The ECL-E-Ns were dissolved in DI water, and shaken vigorously, followed by filtration using a syringe. The filtrate from ECL-E-N (100 μL) was poured into a sterile 96-well plate at various concentrations (0.59–300 mg/mL). Bacterial inoculum (100 μL) in TSB medium with a concentration of 1×10^6 CFU/mL was added. A bacterial suspension with no extracts was employed as a positive control, whereas only extracts in the absence of bacterial suspension was used as a negative control [27]. The 96-well microplates were incubated at 37 °C for 15 h, and then 10 μL of rezasurin solution (0.18% *w*/*v*) was added to each well. The plate was incubated at 37 °C for 2–3 h. The MIC is the lowest concentration (bacteriostatic) of the extract at which there is no color change (blue color) detected. To validate the survival of bacteria, a suspension (10 μL) was taken from each well and spread on TSA agar before being incubated at 37 °C for 24 h. The MIC was the lowest concentration that inhibited the growth of bacteria, whereas the MBC was the lowest concentration of the agent that killed 99.9% of the bacteria or no growth was noted (bactericidal) [28].

2.3.7. Fourier Transform Infrared (FTIR) Spectroscopy

FTIR spectra of the free ECL-E fraction, empty nanoparticles, and ECL-E-loaded nanoparticles (ECL-E-Ns) were determined using an FTIR spectrometer (model Equinox 55, Bruker, Ettlingen, Germany) at a wavenumber between 4000 and 400 cm^{-1}. OPUS software (version 8.5 Bruker Optik GmbH 2020, Bruker, Ettlingen, Germany) was used for analysis.

2.3.8. Scanning Electron Microscopic (SEM) Images of ECL-E-Ns

The selected ECL-E-N samples were visualized using a scanning electron microscope (SEM, FEI Quanta 400-ESEM FEG, Hillsboro, OR, USA). Images were recorded as per the method explained by Sinlapapanya et al. [2].

2.4. Light Stability

The stability of the ECL-E-Ns toward light exposure under varying conditions was tested [3]. An ECL-E-N (10 mL, 50 mg/mL) was placed in a screw-cap test tube. After being exposed to light at varying intensities (1000, 2000, and 4000 lux), samples were taken every day for a total of 6 days for analyses of TPC, ABTS, DPPH, and radical scavenging activities as well as FRAP.

2.5. Statistical Analysis

The experiment was run in triplicate. The SPSS software program (SPSS 22.0 for Windows, SPSS Inc., Chicago, IL, USA) was used to perform an analysis of variance (ANOVA) and Duncan's multiple range test (DMRT) for mean comparisons.

3. Results and Discussion

3.1. Characteristics of ECL-E-Ns Prepared Using CS/ECL-E at Various Ratios by TPP-CS Complexation Methods

3.1.1. Encapsulation Efficiency (EE)

EE indicates the percentage of ECL-E present in nanoparticles. Depending on the CS/ECL-E ratio utilized, the EE of nanoparticles loaded with ECL-E varied from 43.01% to 89.47% (Table 1), suggesting that ECL-E was entrapped in nanoparticles to varied degrees.

CS/ECL-E ratios of 1:2 and 1:3 yielded the highest EE. There was a significant increase in EE ($p < 0.05$) when the CS/ECL-E ratio exceeded 1:1, which might have been caused by more ECL-E being entrapped in nanoparticles. When TPP was added, the positive charge of chitosan could be neutralized by the negative charge of the phosphate, forming a gel that could trap the ECL-E as its core.

EE can be higher with a high volume of extract because of several factors such as (a) increased cross-linking: a higher volume of extract may provide more active compounds that interact with chitosan, leading to better cross-linking with TPP, which stabilizes the nanoparticles and enhances encapsulation [29,30]; (b) higher availability of the encapsulating material: more extract means more material available to be encapsulated, increasing the likelihood of forming more complete nanoparticles with higher payloads [31]; and (c) enhanced viscosity: a greater volume of extract can increase the solution's viscosity, which may slow down particle aggregation and enhance encapsulation by allowing better dispersion of active compounds [32]. These combined phenomena likely contributed to higher EE when using a larger volume of extract in the TPP-CS encapsulation process.

3.1.2. Particle Size (PS), Zeta Potential (ZP), and Polydispersity Index (PDI) of ECL-E-Ns

The PS of different nanoparticles varied with different ratios of ECL-E loaded (Table 1). The PS of ECL-E at varying ratios ranged from 47.30 to 314.60 nm, as shown in Table 1. The result was similar to the study by Al-Nemrawi et al. [33], who found that CS nanoparticle sizes ranged from 145.73 to 724.23 nm. Regardless of the CS/ECL-E ratio, the smallest nanoparticles were found in those devoid of ECL-E, while the largest nanoparticles were found when CS/ECL-E at 1:1 ratio was used ($p < 0.05$). According to Hadidi et al. [22] and Hosseini et al. [34], nanoparticles became larger in size as they were loaded with the core substances. The larger diameter of ECL-E-loaded nanoparticles was mediated by the TPP-CS layer that enveloped the individual ECL-E droplets [34]. As evidenced by a high EE, in the presence of a substantial amount of ECL-E, CS might entrap more ECL-E, especially during the gelation process with the aid of ultrasonication. A smaller nanoparticle could provide the uniform entrapment or encapsulation of ECL-E. Gulzar et al. [7] noticed a similar trend in the particle diameter of CS nanoparticles loaded with shrimp oil. However, there was no difference in the particle sizes of the nanoparticles loaded with ECL-E utilizing the CS/ECL-E ratios of 1:2 and 1:3 ($p > 0.05$).

The PDI is a measure of particle size distribution in a sample. In general, the PDI ranges from 0 (monodisperse particles) to 1 (highly polydisperse particles) [35]. A lower PDI indicates a more uniform particle size distribution, which is often desirable for consistent behavior in applications [33]. ECL-E-N-E2 and ECL-E-N-E3 had the lowest PDI ($p < 0.05$) among all the nanoparticles (0.20–0.22), indicating the most homogeneity or consistency. For CS nanoparticles, a PDI value of less than 0.3 was generally considered acceptable, indicating a relatively narrow size distribution [36]. Strong repulsions among the particles may have resulted from the dispersion of positive charges on the nanoparticles' surfaces, thus reducing the coagulation or collision [23]. Consequently, ECL-E-N-E2 and ECL-E-N-E3 displayed a low PDI.

The zeta potential (ZP) of CS nanoparticles is a critical parameter in determining their stability, interaction with biological systems, and overall functionality [37,38]. ZP is the electrical potential at the slipping plane of a particle in a fluid. It indicates the degree of electrostatic repulsion or attraction between particles [38,39]. All nanoparticles carried positive charges ranging between +40.37 and +44.24 mV, which showed a similar range to the CS nanoparticles prepared by Al-Nemrawi et al. [33]. Adsorption or encapsulation of phenolic extracts in cationic carriers, such as chitosan nanoparticles, typically leads to noticeable changes in the zeta potential of these carriers. This change is primarily due to phenolic compounds, which might carry both negative charge and positively charged surface. During complex formation, the interaction between the phenolic compounds and the cationic carrier could lead to the formation of complexes or conjugates. These complexes might have a lower net positive charge or even shift the zeta potential towards the negative value, depending on the relative amounts and the charge density of the phenolic compounds. A higher zeta potential (either positive or negative) typically reflects better stability of the nanoparticle suspension [40]. Particles with a high zeta potential repel each other and resist aggregation [41]. CS nanoparticles with a high positive zeta potential are more stable in suspension because of electrostatic repulsion [42]. CS nanoparticles often exhibit zeta potentials in the range of +30 to +60 mV, indicating good stability [43].

3.1.3. Color

The L* values of the ECL-E nanoparticles were higher ($p < 0.05$) when compared with those of free ECL-E. This could be the result of the entrapment of ECL-E in the nanoparticles, thus hiding its color [23]. In the TPP-CS encapsulation process, an increased amount of extract encapsulated resulted in a higher L* value (indicating lightness). This was because ECL-E at a higher amount was entrapped, and the surface of wall materials contributed to reflecting lighter and increasing the lightness [44]. In addition, the formation of larger particles or agglomerates could scatter light more effectively, and lighter samples could be noticed [45]. A slightly lower a* value was found in nanoparticles, compared with free ECL-E. Regarding the b* value, similar values were found between ECL-E and nanoparticles loaded with ECL-E at different CS/ECL-E ratios since ECL-E was dechlorophylized before loading into nanoparticles or encapsulation. As a result, green coloring pigments were removed, making the powder fade in color.

3.1.4. In Vitro Releasing Efficiency

Figure 1 shows the releasing efficiency (RE) of nanoparticles at different ECL-E/CS ratios in comparison to the control (1% ECL-E). The nanoparticles showed a slower release rate of ECL-E than that of the control. After 8 h of dialysis, the release profile of 1% ECL-E revealed a roughly 54% RE. RE was significantly decreased by the entrapment of ECL-E in TPP-CS gel. After 8 h, the RE of ECL-E-N-E3, ECL-E-N-E2, and ECL-E-N-E1 was 17%, 20%, and 26%, respectively. ECL-E-N-E3 exhibited the lowest release rate among all the nanoparticles after 72 h. A RE of 51% was reached for ECL-E-N-E3, followed by ECL-E-N-E2 (57%) and ECL-E-N-E1 (63%). The different ECL-E ratios acted as a mediating factor in the variation in their release rates. The reduced RE in the nanoparticle loaded with ECL-E showed that the release of ECL-E was retarded. The releasing efficiency of TPP-CS gel can be influenced by several factors. High cross-linking with TPP can lead to a slow release rate [29]. Small particles often release encapsulated compounds more quickly because of a larger surface area [46,47]. The pH sensitivity of chitosan can affect the swelling of the matrix and, consequently determine the release rate [48]. Controlled release is one of the most desired characteristics of encapsulated substances [24]. Encapsulating ECL-E into nanoparticles significantly reduced the diffusion rate, depending on the amount of ECL-E employed. Furthermore, the manufacturing method controlled their release at different rates.

3.1.5. Antioxidant Activities of TPP-CS Nanoparticles Prepared Using Different CS/ECL-E Ratios

The empty nanoparticles (without loaded ECL-E) and the nanoparticles loaded with ECL-E at varying CS/ECL-E ratios had different antioxidant activities ($p < 0.05$). The amounts of ECL-E corresponding to those entrapped in nanoparticles were used to determine antioxidant activities (Table 2). When the quantity of ECL-E entrapped in the nanoparticles was compared, a higher amount of TPC was correlated with the stronger antioxidant activity of the nanoparticles made with a higher amount of ECL-E. This result revealed that ECL-E was loaded into the nanoparticles, and the ratios affected the amount of ECL-E loaded. Table 2 shows that the produced nanoparticles could significantly increase DPPH radical scavenging, which indicated that encapsulation improved the dispersion of hydrophobic antioxidative compounds [49]. This suggested that the active ingredients in ECL-E might scavenge free radicals that were both amphiphilic (ABTS) and hydrophobic (DPPH) [23,25]. Furthermore, the FRAP assay revealed that ECL-E encapsulated in nanoparticles might provide electrons for free radicals. Additionally, ECL-E also had metal chelating activity, indicating the binding or sequestration of metal ions, prooxidants in the process of lipid oxidation. The ORAC varied among nanoparticles made with varying ECL-E/CS ratios. The highest ORAC was found in the nanoparticles loaded with the highest ratio of ECL-E/CS. This was related to the highest EE or the active compounds loaded as the core of nanoparticles.

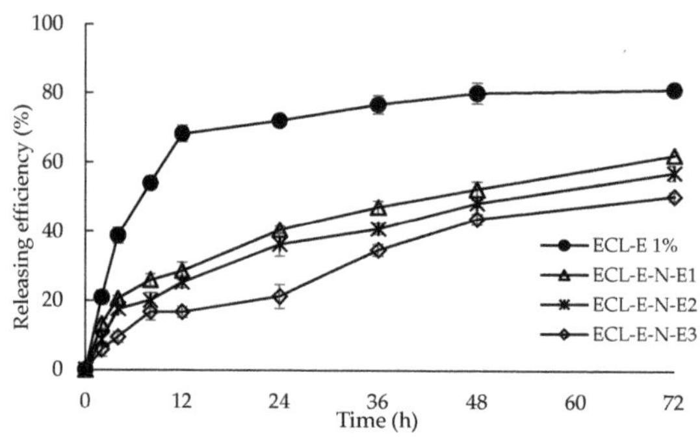

Figure 1. Release profiles of TPP-CS loaded with ECL-E at different levels in comparison with non-encapsulated ECL-E at 1% during a 72 h dialysis period at 4 °C. Bars represent standard deviations (n = 3). ECL-E-N-E1, ECL-E-N-E2, and ECL-E-N-E3 denote nanoparticles using CS/ECL-E ratios of 1:1, 1:2, and 1:3, respectively. ECL-E 1%: free extract at 1%.

Table 2. TPC and antioxidant activities of TPP-CS loaded with ECL-E at different levels in comparison with non-encapsulated ECL-E at 1%.

Treatment	TPC (mg GAE/g Dry Solid)	ABTS (μmol TE/g Dry Solid)	DPPH (mg TE/g dry Solid)	FRAP (mg TE/g Dry Solid)	MCA (μmol EE/g Dry Solid)	ORAC (μmol TE/g Dry Extract)
ECL-E	345.81 ± 16.22 [a]	1409.09 ± 24.26 [a]	362.42 ± 22.24 [c]	892.25 ± 36.87 [a]	4.51 ± 0.14 [c]	73.91 ± 8.67 [b]
TPP-CS	0.33 ± 0.07 [e]	2.15 ± 0.06 [e]	29.01 ± 0.57 [e]	0.08 ± 0.02 [e]	5.81 ± 0.41 [b]	120.75 ± 2.26 [a]
ECL-E-N-E1	47.41 ± 0.74 [d]	438.18 ± 0.83 [d]	68.81 ± 0.24 [d]	156.37 ± 0.35 [d]	6.83 ± 0.38 [a]	26.99 ± 1.44 [d]
ECL-E-N-E2	97.55 ± 0.35 [c]	1302.86 ± 12.06 [b]	1058.81 ± 1.61 [b]	394.20 ± 0.62 [c]	3.30 ± 0.31 [d]	28.93 ± 1.42 [d]
ECL-E-N-E3	253.34 ± 0.20 [b]	1240.80 ± 13.02 [c]	1112.75 ± 13.89 [a]	518.56 ± 0.38 [b]	4.73 ± 0.32 [c]	50.08 ± 1.31 [c]

CS: chitosan, TPP: sodium tripolyphosphate, ECL-E: ethanolic cashew leaf extract. Different lowercase superscripts within the same column indicate significant differences ($p < 0.05$). Mean value and SD± (n = 3).

For the TPP-CS nanoparticles or empty nanoparticles, the antioxidant activities were also evaluated, as shown in Table 2. Because the hydroxyl and amino groups in the chitosan backbone can react with free radicals, empty nanoparticles have the ability to exhibit antioxidant properties. These findings are in line with earlier research conducted by Soleymanfallah et al. [50], especially for DPPH radical scavenging activity and ORAC. For DPPH radical scavenging activity, amino (-NH_2) and hydroxyl (-OH) groups in the structure of chitosan could donate electrons or hydrogen atoms to DPPH free radicals, making DPPH free radicals stable and non-reactive [51]. At the same time, it can donate electrons or hydrogen atoms to oxygen radicals, stabilizing free radicals and reducing oxidative damage [52]. Chitosan has a high ORAC activity. Chitosan with high degree of deacetylation with more free amino groups had, the increases antioxidant capacity [53].

The nanoparticles possessed lower antioxidant activity compared with the free form. Active compounds might bind and disperse within the nanoparticle matrix, potentially lowering their immediate release into solution compared with free compounds [54–56]. When active compounds were encapsulated effectively, the amount available for antioxidant activity could be reduced [55]. A low controlled release can initially reduce the concentration of active compounds available for immediate reaction with radicals [56], which is beneficial for sustained activity.

3.1.6. Antimicrobial Activity

Different MICs and MBCs (Table 3) were found between nanoparticles loaded with ECL-E and free ECL-E ($p < 0.05$). As shown in Table 3, TPP-CS nanoparticles or empty nanoparticles showed antibacterial activity, which was attributed to chitosan's amino group ($-NH_3^+$, positive charge) binding to the bacterial cell wall (negative charge), disrupting the integrity of the cell wall and cell membrane, and possibly causing membrane damage, leakage, and cell death [57,58]. The MICs and MBCs of nanoparticles were higher than those of free ECL-E against *P. aeruginosa* and *S. putrefaciens* ($p < 0.05$), indicating that ECL-E with antimicrobial activity was entrapped within nanoparticles and the release of ECL-E was lowered. This resulted in a decrease in antibacterial activity by initially reducing the concentration of active compounds available for immediate reaction with microorganisms and delaying their antimicrobial effects [59]. Meanwhile, the active agents were also distributed throughout the nanoparticle matrix [55], in which their concentration was diluted compared with the free form. In addition, the barrier from encapsulation was an important factor, potentially limiting their immediate availability [60]. These factors could collectively contribute to decreased antimicrobial activity in nanoparticle formulations compared with free forms.

Table 3. MIC and MBC values of TPP-CS loaded with ECL-E at different levels in comparison with non-encapsulated ECL-E at 1%.

Treatment	*P. aeruginosa* (mg/mL)		*S. putrefaciens* (mg/mL)	
	MIC	MBC	MIC	MBC
ECL-E	2.34 ^{a}	9.38 ^{a}	1.17 ^{a}	4.69 ^{a}
TPP-CS	37.50 ^{d}	300.00 ^{d}	18.75 ^{d}	150.00 ^{c}
ECL-E-N-E1	18.75 ^{c}	150.00 ^{c}	9.38 ^{c}	75.00 ^{b}
ECL-E-N-E2	9.38 ^{b}	75.00 ^{b}	9.38 ^{c}	75.00 ^{b}
ECL-E-N-E3	9.38 ^{b}	75.00 ^{b}	4.69 ^{b}	75.00 ^{b}

CS: chitosan, TPP: sodium tripolyphosphate, ECL-E: ethanolic cashew leaf extract. Different lowercase superscripts within the same column indicate significant differences ($p < 0.05$). Mean value ($n = 3$).

3.1.7. Fourier Transform Infrared Spectroscopic Spectra

The relationship among TPP, CS, and ECL-E in the nanoparticles was verified by FTIR spectroscopy. Figure 2A shows the FTIR spectra of ECL-E and nanoparticles without and with ECL-E. The peaks at 3600–2800 cm^{-1} and 1700–400 cm^{-1} were the main peaks in the infrared spectra. This was similar to the study by Antoniou et al. [61]. The peaks at 3400–3200 cm^{-1} were a result of the −OH stretching vibration associated with free, inter-, and intramolecular bound hydroxyl groups for ECL-E [61]. In addition, these peaks might also be associated with hydroxyl groups in free chitosan and CS-TPP cross-links in nanoparticles [7]. In the same area, the peaks at ≈3300 cm^{-1} in nanoparticles were attributed to N−H (amino) group stretching [7]. Peaks in the range of 3000–2800 cm^{-1} corresponded to the typical C−H vibration [61]. The asymmetric stretching of the C−O−C bond in the glycosidic linkage appeared around 1150 cm^{-1}, while the peak at 1077 cm^{-1} corresponded to the C−O stretching vibration [61]. The peaks in the range of 1220–1080 cm^{-1} were attributed to P=O stretching, while the absorption bands at 843 cm^{-1} corresponded to P−O and P−O−P in the TPP spectrum [61]. Additionally, the peak at 1204 cm^{-1} in the nanoparticle spectra indicated P=O stretching, resulting from the cross-linking between CS and TPP [62]. The bioactive chemicals in ECL-E exhibited strong peaks at the following wavenumbers: 3383 cm^{-1}, which corresponded to O–H stretching; 1544 cm^{-1}, which represented C=C stretching; 1447 cm^{-1}, which corresponded to C–C stretching; and 1037 cm^{-1}, which indicated C–O–C stretching (Figure 2) [63]. A portion of these peaks were found in ECL-E-Ns. Finally, the peak for N−H bending vibration of primary amines at 1639 cm^{-1} in the nanoparticle spectra shifted to 1626 cm^{-1} when loaded with ECL-E in ECL-E-Ns. These results indicated the interaction between ECL-E and nanoparticles. The emergence of these peaks signifies the production of nanoparticles and the enhancement of inter- and intramolecular interactions in nanoparticles.

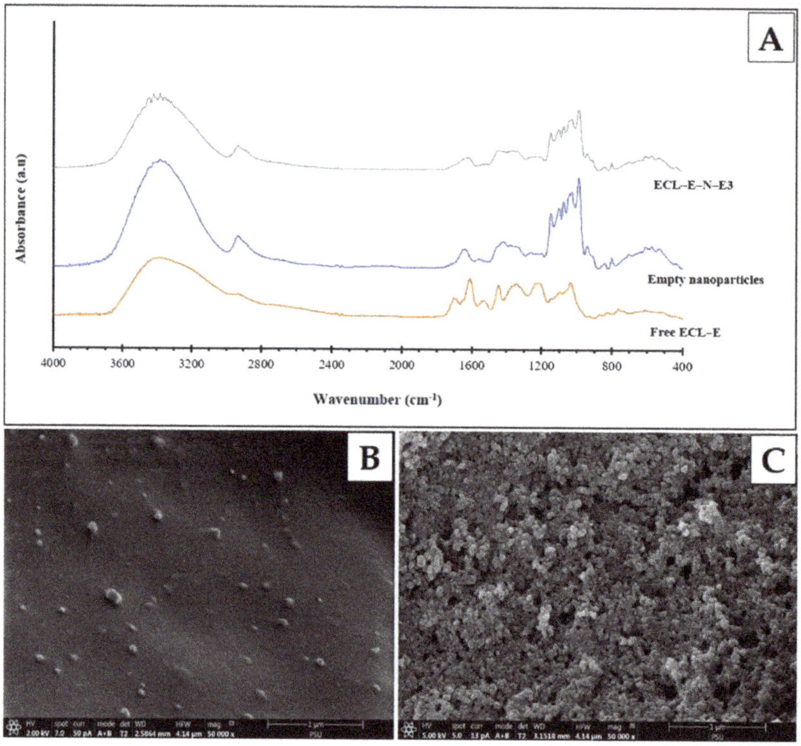

Figure 2. FTIR spectra of free ECL-E, empty nanoparticles, and ECL-E-N-E3 (**A**). Scanning electron micrographs of empty nanoparticles (**B**) and ECL-E-N-E3 (**C**). ECL-E-N-E3: nanoparticles using CS/ECL-E ratios of 1:3; Free ECL-E: free extract at 1%.

3.1.8. Scanning Electron Microscopic Images

The nanoparticles were spherical and nanometer in size, as displayed in Figure 2B,C. In Figure 2C, the nanoparticles loaded with ECL-E have a continuous and regular shape. Their morphology could be affected by the freeze-drying method. During drying, particles align closer together and bind together [64,65]. Figure 2B represents nanoparticles without ECL-E loaded, which had an uneven distribution because of the fact that the nanoparticles without ECL-E loaded did not have the core of the extract [65,66]. The particle size was thus smaller than nanoparticles loaded with ECL-E, as shown in Table 1. Figure 2C shows that the nanoparticles loaded with ECL-E have larger particle diameters (Table 1). The nanoparticles in each treatment had varying particle sizes. Size could be governed by the electrostatic interaction of TPP and CS in the ionic gelation process and was dependent on the mixing occurring during preparation [7].

3.2. Light Stability of TPP-CS Nanoparticles Loaded with ECL-E as Affected by Light at Various Intensities

Figure 3 illustrates the TPC, ABTS, and DPPH radical scavenging activities and FRAP of nanoparticles loaded with ECL-E. The TPC and antioxidant activities of nanoparticles exposed to light increased in the first 48 h and then progressively declined after 72 h. The increased light intensity caused the rapid degradation, in which some products with antioxidant properties were produced. Esua et al. [67] also found a similar result, in which the antioxidant of tomato dried for a longer time had increased antioxidant activity. As shown in Figure 3, the period of exposure was another significant factor that influenced the breakdown of phenolic compounds. For the longest duration (144 h), the nanoparticles were exposed to the maximum intensity of light (4000 lux). This resulted in the greatest losses in TPC and, naturally antioxidant activities.

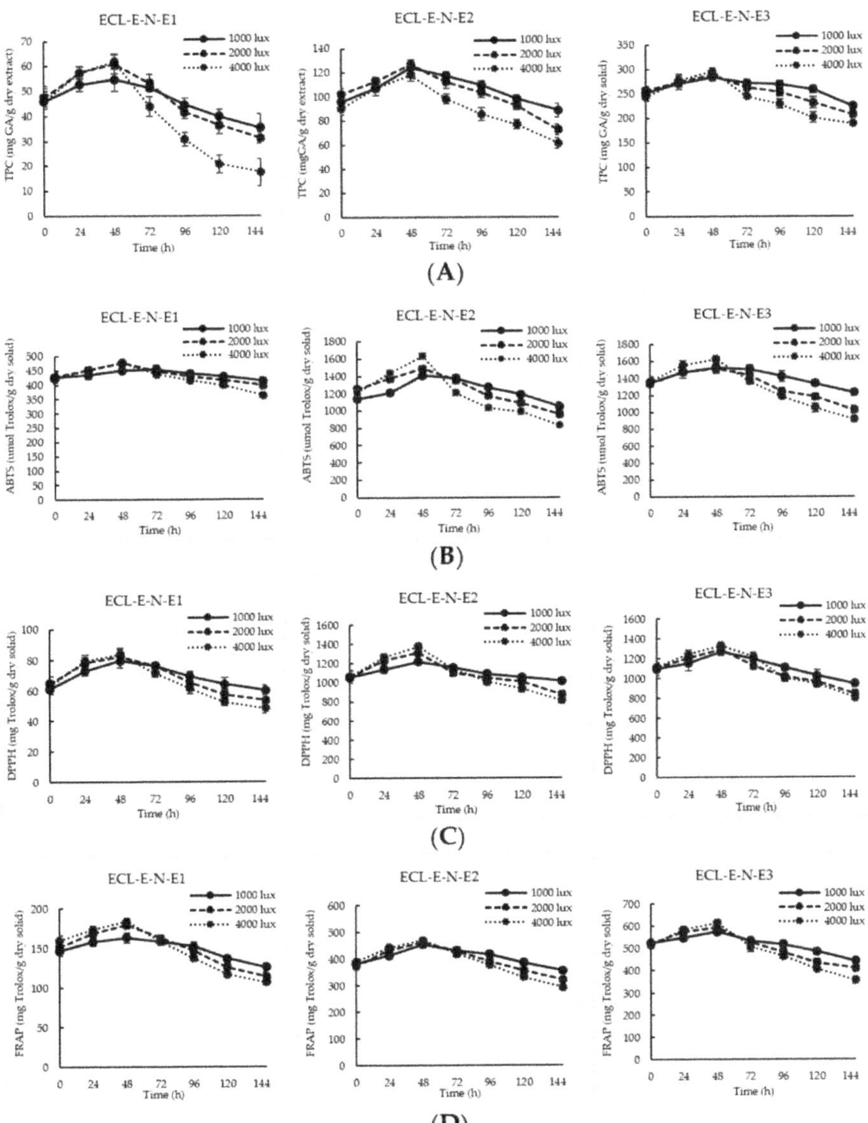

Figure 3. Effect of light at different intensities and times on the total phenolic content (TPC, (**A**)), ABTS radical scavenging activity (ABTS, (**B**)), DPPH radical scavenging activity (DPPH, (**C**)), and ferric reducing antioxidant power (FRAP, (**D**)) of TPP-CS loaded with ECL-E at different levels. Bars represent standard deviations (n = 3). ECL-E-N-E1, ECL-E-N-E2, and ECL-E-N-E3 denote nanoparticles using CS/ECL-E ratios of 1:1, 1:2, and 1:3, respectively.

Light intensity can affect nanoparticles by causing degradation or structural changes, especially UV, which can degrade chitosan, breaking down its molecular structure [12,68,69]. At the same time, high-intensity light can increase the temperature, potentially affecting the stability of the nanoparticles [70]. These effects can reduce the effectiveness and stability of chitosan nanoparticles. This was evidenced by an increase in TPC and antioxidant activities. Because of the damage to nanoparticles, the ECL-E inside was released. When the ECL-E was released freely, polyphenols with several OH groups could possess a high capacity as antioxidants. However, those compounds were mostly light-sensitive [71]. Therefore, it decomposes and loses its antioxidant activities easily. Nevertheless, those compounds could be protected or the loss of bioactive compounds in ECL-E could be delayed to a certain extent when compared with free ECL-E (un-encapsulation) [3]. This result was also in accordance with the study on cashew leaf extract by Sinlapapanya et al. [3], who documented that TPC decreased when exposed to light and decreased to a higher extent when exposed to high-intensity light.

4. Conclusions

Encapsulation of ECL-E with TPP-CS nanoparticles by the ionic gelation method was carried out. The ECL-E-N-E3 nanoparticles exhibited excellent physical properties and bioactivities. It had a pale color, tiny particle size, positive zeta potential, and low PDI with high antioxidant and antibacterial activities. The ECL-E/CS ratio had a significant impact on the interaction among TPP, CS, and ECL-E for the creation of nanoparticles. Furthermore, it also affected the shape of nanoparticles. The nanoparticles demonstrated a protective effect against the loss of ECL-E activity induced by light intensity and also played a significant function in regulating the rate of ECL-E release. Thus, ECL-E encapsulated with TPP-CS nanoparticles could serve as a delivery vehicle with enhanced stability by maintaining the antioxidant activity and antibacterial activity of ECL-E.

Author Contributions: Conceptualization, P.S., J.B. and S.B.; methodology, P.S. and J.B.; validation, P.S., J.B. and S.B.; formal analysis, P.S.; investigation, P.S. and S.B.; resources, S.B.; data curation, P.S.; writing—original draft preparation, P.S.; writing—review and editing, S.P., R.A.N., B.Z., T.P. and S.B.; supervision, S.B.; project administration, S.B.; funding acquisition, S.B. All authors have read and agreed to the published version of the manuscript.

Funding: This study was funded by a Prachayacharn grant from Prince of Songkla University (Grant No. AGR6602079N) and the National Research Council of Thailand (NRCT) under the International Research Network (IRN) program.

Data Availability Statement: The original contributions presented in the study are included in the article, further inquiries can be directed to the corresponding author.

Acknowledgments: The authors would like to thank the International Center of Excellence in Seafood Science and Innovation, Faculty of Agro-Industry, Prince of Songkla University for the facilities provided for all the experiments.

Conflicts of Interest: The authors declare no conflicts of interest.

References

1. Pinto, T.; Aires, A.; Cosme, F.; Bacelar, E.; Morais, M.C.; Oliveira, I.; Ferreira-Cardoso, J.; Anjos, R.; Vilela, A.; Gonçalves, B. Bioactive (poly) Phenols, Volatile Compounds from Vegetables, Medicinal and Aromatic Plants. *Foods* **2021**, *10*, 106. [CrossRef] [PubMed]
2. Sinlapapanya, P.; Sumpavapol, P.; Nirmal, N.; Zhang, B.; Hong, H.; Benjakul, S. Ethanolic Cashew Leaf Extract: Antimicrobial Activity, Mode of Action, and Retardation of Spoilage Bacteria in Refrigerated Nile Tilapia Slices. *Foods* **2022**, *11*, 3461. [CrossRef]
3. Sinlapapanya, P.; Sumpavapol, P.; Buatong, J.; Benjakul, S. Ethanolic Cashew Leaf Extract: Antioxidant Potential and Impact on Quality Changes of Dried Fermented Catfish During Storage. *Future Foods* **2024**, *9*, 100296. [CrossRef]
4. Cao, H.; Saroglu, O.; Karadag, A.; Diaconeasa, Z.; Zoccatelli, G.; Conte-Junior, C.A.; Gonzalez-Aguilar, G.A.; Ou, J.; Bai, W.; Zamarioli, C.M.; et al. Available Technologies on Improving The Stability of Polyphenols in Food Processing. *Food Front.* **2021**, *2*, 109–139. [CrossRef]

5. Yin, Z.; Zheng, T.; Ho, C.T.; Huang, Q.; Wu, Q.; Zhang, M. Improving The Stability and Bioavailability of Tea Polyphenols by Encapsulations: A review. *Food Sci. Hum. Wellness* **2022**, *11*, 537–556. [CrossRef]
6. Assadpour, E.; Jafari, S.M. Advances in Spray-Drying Encapsulation of Food Bioactive Ingredients: From Microcapsules to Nanocapsules. *Annu. Rev. Food Sci. Technol.* **2019**, *10*, 103–131. [CrossRef]
7. Gulzar, S.; Raju, N.; Prodpran, T.; Benjakul, S. Chitosan-Tripolyphosphate Nanoparticles Improves Oxidative Stability of Encapsulated Shrimp Oil Throughout The Extended Storage. *Eur. J. Lipid Sci. Technol.* **2022**, *124*, 2100178. [CrossRef]
8. Maleki, G.; Woltering, E.J.; Mozafari, M.R. Applications of Chitosan-Based Carrier as an Encapsulating Agent in Food Industry. *Trends Food Sci. Technol.* **2022**, *120*, 88–99. [CrossRef]
9. Akhavan, S.; Assadpour, E.; Katouzian, I.; Jafari, S.M. Lipid Nano Scale Cargos for The Protection and Delivery of Food Bioactive Ingredients and Nutraceuticals. *Trends Food Sci. Technol.* **2018**, *74*, 132–146. [CrossRef]
10. Rostami, M.; Yousefi, M.; Khezerlou, A.; Mohammadi, M.A.; Jafari, S.M. Application of Different Biopolymers for Nanoencapsulation of Antioxidants via Electrohydrodynamic Processes. *Food Hydrocoll.* **2019**, *97*, 105170. [CrossRef]
11. Jacob, P.L.; Brugnoli, B.; Del Giudice, A.; Phan, H.; Chauhan, V.M.; Beckett, L.; Gillis, R.B.; Moloney, C.; Cavanagh, R.J.; Krumins, E.; et al. Poly (diglycerol adipate) Variants as Enhanced Nanocarrier Replacements in Drug Delivery Applications. *J. Colloid Interface Sci.* **2023**, *641*, 1043–1057. [CrossRef] [PubMed]
12. Wang, W.; Xue, C.; Mao, X. Chitosan: Structural Modification, Biological Activity and Application. *Int. J. Biol. Macromol.* **2020**, *164*, 4532–4546. [CrossRef] [PubMed]
13. Seidi, F.; Yazdi, M.K.; Jouyandeh, M.; Dominic, M.; Naeim, H.; Nezhad, M.N.; Bagheri, B.; Habibzadeh, S.; Zarrintaj, P.; Saeb, M.R.; et al. Chitosan-Based Blends for Biomedical Applications. *Int. J. Biol. Macromol.* **2021**, *183*, 1818–1850. [CrossRef] [PubMed]
14. Kumari, S.; Kishor, R. Chapter 1-Chitin and Chitosan: Origin, Properties, and Applications. In *Handbook of Chitin and Chitosan;* Gopi, S., Thomas, S., Pius, A., Eds.; Elsevier: Amsterdam, The Netherlands, 2020; pp. 1–33.
15. Pellis, A.; Guebitz, G.M.; Nyanhongo, G.S. Chitosan: Sources, Processing and Modification Techniques. *Gels* **2022**, *8*, 393. [CrossRef] [PubMed]
16. Brugnoli, B.; Mariano, A.; Simonis, B.; Bombelli, C.; Sennato, S.; Piozzi, A.; Taresco, V.; Chauhan, V.M.; Howdle, S.M.; d'Abusco, A.S.; et al. Self-assembled Chitosan-sodium Caseinate Drug Delivery Nanosystems: Synthesis, Characterization, Stability Studies, in vitro Cytotoxicity and in vivo Biocompatibility against 143 B Cells. *Carbohydr. Polym. Technol. Appl.* **2023**, *6*, 100373. [CrossRef]
17. Shafiei, M.; Jafarizadeh-Malmiri, H.; Rezaei, M. Biological Activities of Chitosan and Prepared Chitosan-Tripolyphosphate Nanoparticles using Ionic Gelation Method Against Various Pathogenic Bacteria and Fungi Strains. *Biologia* **2019**, *74*, 1561–1568. [CrossRef]
18. Gutiérrez-Ruíz, S.C.; Cortes, H.; González-Torres, M.; Almarhoon, Z.M.; Gürer, E.S.; Sharifi-Rad, J.; Leyva-Gómez, G. Optimize The Parameters for The Synthesis By The Ionic Gelation Technique, Purification, and Freeze-Drying of Chitosan-Sodium Tripolyphosphate Nanoparticles for Biomedical Purposes. *J. Biol. Eng.* **2024**, *18*, 12. [CrossRef]
19. Akbari-Alavijeh, S.; Shaddel, R.; Jafari, S.M. Encapsulation of Food Bioactives and Nutraceuticals By Various Chitosan-Based Nanocarriers. *Food Hydrocoll.* **2020**, *105*, 105774. [CrossRef]
20. Di Santo, M.C.; D'Antoni, C.L.; Rubio, A.P.D.; Alaimo, A.; Pérez, O.E. Chitosan-Tripolyphosphate Nanoparticles Designed to Encapsulate Polyphenolic Compounds for Biomedical and Pharmaceutical Applications—A review. *Biomed. Pharmacother.* **2021**, *142*, 111970. [CrossRef] [PubMed]
21. Shetta, A.; Kegere, J.; Mamdouh, W. Comparative Study of Encapsulated Peppermint and Green Tea Essential Oils in Chitosan Nanoparticles: Encapsulation, Thermal Stability, In-vitro Release, Antioxidant and Antibacterial Activities. *Int. J. Biol. Macromol.* **2019**, *126*, 731–742. [CrossRef] [PubMed]
22. Hadidi, M.; Pouramin, S.; Adinepour, F.; Haghani, S.; Jafari, S.M. Chitosan Nanoparticles Loaded with Clove Essential Oil: Characterization, Antioxidant and Antibacterial Activities. *Carbohydr. Polym.* **2020**, *236*, 116075. [CrossRef] [PubMed]
23. Tagrida, M.; Prodpran, T.; Zhang, B.; Aluko, R.E.; Benjakul, S. Liposomes Loaded with Betel Leaf (*Piper betle* L.) Ethanolic Extract Prepared by Thin Film Hydration and Ethanol Injection Methods: Characteristics and Antioxidant Activities. *J. Food Biochem.* **2021**, *45*, e14012. [CrossRef] [PubMed]
24. Olatunde, O.O.; Benjakul, S.; Vongkamjan, K.; Amnuaikit, T. Liposomal Encapsulated Ethanolic Coconut Husk Extract: AntiOxidant and Antibacterial Properties. *J. Food Sci.* **2019**, *84*, 3664–3673. [CrossRef] [PubMed]
25. Benjakul, S.; Kittiphattanabawon, P.; Sumpavapol, P.; Maqsood, S. Antioxidant Activities of Lead (*Leucaena leucocephala*) Seed as Affected By Extraction Solvent, Prior Dechlorophyllisation and Drying Methods. *J. Food Sci. Technol.* **2014**, *51*, 3026–3037. [CrossRef]
26. Sae-Leaw, T.; O'Callaghan, Y.C.; Benjakul, S.; O'Brien, N.M. Antioxidant Activities and Selected Characteristics of Gelatin Hydrolysates from Seabass (*Lates calcarifer*) Skin as Affected by Production Processes. *J. Food Sci. Technol.* **2016**, *53*, 197–208. [CrossRef] [PubMed]
27. Lopez, M.K.N.; Hadisurya, M.; Cornwall, R.G. Antimicrobial Investigation and Structure Activity Analysis of Natural Eugenol Derivatives Against Several Oral Bacteria. *J. Pharm. Microbiol.* **2019**, *5*, 1–4.
28. Jafarı-sales, A.; Jafari, B.; Khaneshpour, H.; Pashazadeh, M. Antibacterial Effect of Methanolic Extract of Rosa Damascena on Standard Bacteria *Staphylococcus aureus*, *Bacillus cereus*, *Escherichia coli* and *Pseudomonas aeruginosa* In Vitro. *Int. J. Nat. Life Sci.* **2020**, *4*, 40–46.

29. Kadam, D.; Lele, S.S. Cross-Linking Effect of Polyphenolic Extracts of *Lepidium sativum* Seedcake on Physicochemical Properties of Chitosan Films. *Int. J. Biol. Macromol.* **2018**, *114*, 1240–1247. [CrossRef]
30. Wahba, M.I. Enhancement of The Mechanical Properties of Chitosan. *J. Biomater. Sci. Polym. Ed.* **2020**, *31*, 350–375. [CrossRef]
31. Da Rosa, C.G.; Borges, C.D.; Zambiazi, R.C.; Rutz, J.K.; da Luz, S.R.; Krumreich, F.D.; Benvenutti, E.V.; Nunes, M.R. Encapsulation of The Phenolic Compounds of The Blackberry (*Rubus fruticosus*). *LWT-Food Sci. Technol.* **2014**, *58*, 527–533. [CrossRef]
32. Kaushalya, K.G.D.; Gunathilake, K.D.P.P. Encapsulation of Phlorotannins From Edible Brown Seaweed in Chitosan: Effect of Fortification on Bioactivity and Stability in Functional Foods. *Food Chem.* **2022**, *377*, 132012. [CrossRef] [PubMed]
33. Al-Nemrawi, N.K.; Alsharif, S.S.M.; Dave, R.H. Preparation of Chitosan-TPP Nanoparticles: The Influence of Chitosan Polymeric Properties and Formulation Variables. *Int. J. Appl. Pharm.* **2018**, *10*, 60–65. [CrossRef]
34. Hosseini, S.F.; Zandi, M.; Rezaei, M.; Farahmandghavi, F. Two-Step Method for Encapsulation of Oregano Essential Oil in Chitosan Nanoparticles: Preparation, Characterization and In Vitro Release Study. *Carbohydr. Polym.* **2013**, *95*, 50–56. [CrossRef]
35. Danaei, M.R.M.M.; Dehghankhold, M.; Ataei, S.; Hasanzadeh Davarani, F.; Javanmard, R.; Dokhani, A.; Khorasani, S.; Mozafari, M.R. Impact of Particle Size and Polydispersity Index on The Clinical Applications of Lipidic Nanocarrier Systems. *Pharmaceutics* **2018**, *10*, 57. [CrossRef] [PubMed]
36. Yan, Q.; Weng, J.; Wu, X.; Wang, W.; Yang, Q.; Guo, F.; Wu, D.; Song, Y.; Chen, F.; Yang, G. Characteristics, Cryoprotection Evaluation and In Vitro Release of BSA-Loaded Chitosan Nanoparticles. *Mar. Drugs* **2020**, *18*, 315. [CrossRef]
37. Ferreira, L.M.; Dos Santos, A.M.; Boni, F.I.; Dos Santos, K.C.; Robusti, L.M.G.; de Souza, M.P.; Ferreira, N.N.; Carvalho, S.G.; Cardoso, V.M.O.; Chorilli, M.; et al. Design of Chitosan-Lased Particle Systems: A Review of The Physicochemical Foundations for Tailored Properties. *Carbohydr. Polym.* **2020**, *250*, 116968. [CrossRef]
38. Cano-Sarmiento, C.T.D.I.; Téllez-Medina, D.I.; Viveros-Contreras, R.; Cornejo-Mazón, M.; Figueroa-Hernández, C.Y.; García-Armenta, E.; Alamilla-Beltrán, L.; García, H.S.; Gutiérrez-López, G.F. Zeta Potential of Food Matrices. *Food Eng. Rev.* **2018**, *10*, 113–138. [CrossRef]
39. Prakash, S.; Mishra, R.; Malviya, R.; Sharma, P.K. Measurement Techniques and Pharmaceutical Applications of Zeta Potential: A Review. *J. Chronother. Drug Deliv.* **2014**, *5*, 33–40.
40. Cacua, K.; Ordoñez, F.; Zapata, C.; Herrera, B.; Pabón, E.; Buitrago-Sierra, R. Surfactant Concentration and pH Effects on The Zeta Potential Values of Alumina Nanofluids to Inspect Stability. *Colloids Surf. A Physicochem. Eng. Asp.* **2019**, *583*, 123960. [CrossRef]
41. Shrestha, S.; Wang, B.; Dutta, P. Nanoparticle Processing: Understanding and Controlling Aggregation. *Adv. Colloid Interface Sci.* **2020**, *279*, 102162. [CrossRef] [PubMed]
42. Collado-González, M.; Montalbán, M.G.; Peña-García, J.; Pérez-Sánchez, H.; Víllora, G.; Baños, F.G.D. Chitosan as Stabilizing Agent for Negatively Charged Nanoparticles. *Carbohydr. Polym.* **2017**, *161*, 63–70. [CrossRef]
43. Ozturk, K.; Arslan, F.B.; Tavukcuoglu, E.; Esendagli, G.; Calis, S. Aggregation of Chitosan Nanoparticles in Cell Culture: Reasons and Resolutions. *Int. J. Pharm.* **2020**, *578*, 119119. [CrossRef] [PubMed]
44. Haas, K.; Obernberger, J.; Zehetner, E.; Kiesslich, A.; Volkert, M.; Jaeger, H. Impact of Powder Particle Structure on The Oxidation Stability and Color of Encapsulated Crystalline and Emulsified Carotenoids in Carrot Concentrate Powders. *J. Food Eng.* **2019**, *263*, 398–408. [CrossRef]
45. Zubko, E.; Shkuratov, Y.; Videen, G. Effect of Morphology on Light Scattering By Agglomerates. *J. Quant. Spectrosc. Radiat. Transf.* **2015**, *150*, 42–54. [CrossRef]
46. Saifullah, M.; Shishir, M.R.I.; Ferdowsi, R.; Rahman, M.R.T.; Van Vuong, Q. Micro and Nano Encapsulation, Retention and Controlled Release of Flavor and Aroma Compounds: A Critical Review. *Trends Food Sci. Technol.* **2019**, *86*, 230–251. [CrossRef]
47. Jafari, S.M.; He, Y.; Bhandari, B. Role of Powder Particle Size on The Encapsulation Efficiency of Oils During Spray Drying. *Dry. Technol.* **2007**, *25*, 1081–1089. [CrossRef]
48. Rizwan, M.; Yahya, R.; Hassan, A.; Yar, M.; Azzahari, A.D.; Selvanathan, V.; Sonsudin, F.; Abouloula, C.N. pH Sensitive Hydrogels in Drug Delivery: Brief History, Properties, Swelling, and Release Mechanism, Material Selection and Applications. *Polymers* **2017**, *9*, 137. [CrossRef] [PubMed]
49. Shah, B.R.; Zhang, C.; Li, Y.; Li, B. Bioaccessibility and Antioxidant Activity of Curcumin after Encapsulated by Nano and Pickerng Emulsion Based on Chitosan-Tripolyphosphate Nanoparticles. *Food Res. Int.* **2016**, *89*, 399–407. [CrossRef]
50. Soleymanfallah, S.; Khoshkhoo, Z.; Hosseini, S.E.; Azizi, M.H. Preparation, Physical Properties, and Evaluation of Antioxidant Capacity of Aqueous Grape Extract Loaded in Chitosan-TPP Nanoparticles. *Food Sci. Nutr.* **2022**, *10*, 3272–3281. [CrossRef]
51. Abd El-Hack, M.E.; El-Saadony, M.T.; Shafi, M.E.; Zabermawi, N.M.; Arif, M.; Batiha, G.E.; Khafaga, A.F.; El-Hakim, Y.M.A.; Al-Sagheer, A.A. Antimicrobial and Antioxidant Properties of Chitosan and Its Derivatives and Their Applications: A Review. *Int. J. Biol. Macromol.* **2020**, *164*, 2726–2744. [CrossRef]
52. Li, Q.; Mi, Y.; Tan, W.; Guo, Z. Highly Efficient Free Radical-Scavenging Property of Phenolic-Functionalized Chitosan Derivatives: Chemical Modification and Activity Assessment. *Int. J. Biol. Macromol.* **2020**, *164*, 4279–4288. [CrossRef] [PubMed]
53. Hao, W.; Li, K.; Ma, Y.; Li, R.; Xing, R.; Yu, H.; Li, P. Preparation and Antioxidant Activity of Chitosan Dimers with Different Sequences. *Mar. Drugs* **2021**, *19*, 366. [CrossRef] [PubMed]
54. Ge, X.; Cao, Z.; Chu, L. The Antioxidant Effect of The Metal and Metal-Oxide Nanoparticles. *Antioxidants* **2022**, *11*, 791. [CrossRef] [PubMed]
55. Maqsoudlou, A.; Assadpour, E.; Mohebodini, H.; Jafari, S.M. The Influence of Nanodelivery Systems on the Antioxidant Activity of Natural Bioactive Compounds. *Crit. Rev. Food Sci. Nutr.* **2022**, *62*, 3208–3231. [CrossRef]

56. Li, Q.; Duan, M.; Liu, L.; Chen, X.; Fu, Y.; Li, J.; Zhao, T.; McClements, D.J. Impact of Polyphenol Interactions with Titanium dioxide Nanoparticles on Their Bioavailability and Antioxidant Activity. *J. Agric. Food Chem.* **2021**, *69*, 9661–9670. [CrossRef]
57. Yilmaz Atay, H. Antibacterial Activity of Chitosan-Based Systems. In *Functional Chitosan: Drug Delivery and Biomedical Applications*; Jana, S., Jana, S., Eds.; Springer: Singapore, 2019; pp. 457–489.
58. Koilparambil, D.; Varghese, S.; Shaikmoideen, J.M. Chitosan Nanoparticles: A Novel Antimicrobial Agent. In *Nanobiotechnology in Diagnosis, Drug Delivery, and Treatment*; Rai, M., Razzaghi-Abyaneh, M., Ingle, A.P., Eds.; John Wiley and Sons: Chichester, UK, 2020; pp. 197–215.
59. Kuai, L.; Liu, F.; Chiou, B.S.; Avena-Bustillos, R.J.; McHugh, T.H.; Zhong, F. Controlled Release of Antioxidants from Active Food Packaging: A Review. *Food Hydrocoll.* **2021**, *120*, 106992. [CrossRef]
60. Subroto, E.; Andoyo, R.; Indiarto, R. Solid Lipid Nanoparticles: Review of The Current Research on Encapsulation and Delivery Systems for Active and Antioxidant Compounds. *Antioxidants* **2023**, *12*, 633. [CrossRef]
61. Antoniou, J.; Liu, F.; Majeed, H.; Qi, J.; Yokoyama, W.; Zhong, F. Physicochemical and Morphological Properties of Size-Controlled Chitosan–Tripolyphosphate Nanoparticles. *Colloids Surf. A Physicochem. Eng. Asp.* **2015**, *465*, 137–146. [CrossRef]
62. Khoerunnisa, F.; Nurhayati, M.; Dara, F.; Rizki, R.; Nasir, M.; Aziz, H.A.; Hendrawan, H.; Poh, N.E.; Kaewsaneha, C.; Opaprakasit, P. Physicochemical Properties of TPP-Crosslinked Chitosan Nanoparticles as Potential Antibacterial Agents. *Fibers Polym.* **2021**, *22*, 2954–2964. [CrossRef]
63. Lubis, L.D.; Prananda, A.T.; Juwita, N.A.; Nasution, M.A.; Syahputra, R.A.; Sumaiyah, S.; Lubis, R.R.; Lubis, M.F.; Astyka, R.; Atiqah, J.F. Unveiling Antioxidant Capacity of Standardized Chitosan-Tripolyphosphate Microcapsules Containing PolyPhenol-Rich Extract of *Portulaca oleraceae*. *Heliyon* **2024**, *10*, e29541. [CrossRef]
64. Luo, W.C.; Beringhs, A.O.R.; Kim, R.; Zhang, W.; Patel, S.M.; Bogner, R.H.; Lu, X. Impact of Formulation on The Quality and Stability of Freeze-Dried Nanoparticles. *Eur. J. Pharm. Biopharm.* **2021**, *169*, 256–267. [CrossRef] [PubMed]
65. Tchabo, W.; Ma, Y.; Kaptso, G.K.; Kwaw, E.; Cheno, R.W.; Wu, M.; Osae, R.; Ma, S.; Farooq, M. Carrier Effects on The Chemical and Physical Properties of Freeze-Dried Encapsulated Mulberry Leaf Extract Powder. *Acta Chim. Slov.* **2018**, *65*, 823–835. [CrossRef] [PubMed]
66. Wais, U.; Jackson, A.W.; He, T.; Zhang, H. Nanoformulation and Encapsulation Approaches for Poorly Water-Soluble Drug Nanoparticles. *Nanoscale* **2016**, *8*, 1746–1769. [CrossRef] [PubMed]
67. Esua, O.J.; Chin, N.L.; Yusof, Y.A.; Sukor, R. Effects of Simultaneous UV-C Radiation and Ultrasonic Energy Postharvest Treatment on Bioactive Compounds and Antioxidant Activity of Tomatoes During Storage. *Food Chem.* **2019**, *270*, 113–122. [CrossRef]
68. Yue, W.; Yao, P.; Wei, Y.; Mo, H. Synergetic Effect of Ozone and Ultrasonic Radiation on Degradation of Chitosan. *Polym. Degrad. Stab.* **2008**, *93*, 1814–1821. [CrossRef]
69. Bussiere, P.O.; Gardette, J.L.; Rapp, G.; Masson, C.; Therias, S. New Insights into The Mechanism of Photodegradation of Chitosan. *Carbohydr. Polym.* **2021**, *259*, 117715. [CrossRef]
70. Harish, V.; Ansari, M.M.; Tewari, D.; Yadav, A.B.; Sharma, N.; Bawarig, S.; García-Betancourt, M.L.; Karatutlu, A.; Bechelany, M.; Barhoum, A. Cutting-Edge Advances in Tailoring Size, Shape, and Functionality of Nanoparticles and Nanostructures: A review. *J. Taiwan Inst. Chem. Eng.* **2023**, *149*, 105010. [CrossRef]
71. Esparza, I.; Cimminelli, M.J.; Moler, J.A.; Jiménez-Moreno, N.; Ancín-Azpilicueta, C. Stability of Phenolic Compounds in Grape Stem Extracts. *Antioxidants* **2020**, *9*, 720. [CrossRef]

colloids and interfaces

MDPI

Article

Encapsulation of Canola Oil by Sonication for the Development of Protein and Starch Systems: Physical Characteristics and Rheological Properties

Reynaldo J. Silva-Paz [1,*], Celenia E. Ñope-Quito [1], Thalia A. Rivera-Ashqui [2], Nicodemo C. Jamanca-Gonzales [1], Amparo Eccoña-Sota [3], Natalia Riquelme [4] and Carla Arancibia [4]

[1] Departamento de Ingeniería, Escuela de Ingeniería en Industrias Alimentarias, Universidad Nacional de Barranca, Av. Toribio de Luzuriaga N° 376 Mz J-Urbanización La Florida, Barranca 15169, Peru; cnopeq162@unab.edu.pe (C.E.Ñ.-Q.); njamanca@unab.edu.pe (N.C.J.-G.)
[2] CONICET—Consejo Nacional de Investigaciones Científicas y Técnicas, Godoy Cruz 2290, Buenos Aires C1425FQB, Argentina; thalia.rivera@conicet.gov.ar
[3] Facultad de Medicina Veterinaria y Zootecnia, Universidad Nacional Micaela Bastidas de Apurímac, Av. Inca Garcilaso de la Vega s/n, Tamburco, Abancay, Apurímac 03001, Peru; aecconha@unamba.edu.pe
[4] Department of Food Science and Technology, Universidad de Santiago de Chile, Obispo Umaña 050, Estación Central 9170201, Chile; natalia.riquelme.h@usach.cl (N.R.); carla.arancibia@usach.cl (C.A.)
* Correspondence: rsilva@unab.edu.pe; Tel.: +51-957805238

Abstract: Canola oil, extracted from *Brassica napus*, is appreciated for its nutritional profile, but its use in the food industry is limited by its susceptibility to oxidation. This study aimed to evaluate the nanoemulsion of canola oil by sonication to develop stable nanoemulsified gels from protein and starch systems. Two stages were performed. In the first stage, oil-in-water (O/W) nanoemulsions were prepared using soy lecithin and Tween 80 as emulsifiers, analyzing their physical stability by particle size and polydispersity index. The results show that the sonication conditions and emulsifier concentration significantly affected the creaming index and particle size. In the second stage, gels were developed from these nanoemulsions, evaluating their colorimetric and rheological properties. It was observed that the gels presented a viscoelastic behavior suitable for food applications, with a higher luminosity in protein systems. In conclusion, nanoemulsion by sonication improves the stability of canola oil, suggesting its potential use in various food applications. Additional emulsifier combinations and optimization of processing conditions are recommended to further improve the stability and functionality of the encapsulated oil.

Keywords: encapsulation; color; canola; optimization; rheology

1. Introduction

Canola oil, extracted from the Brassica napus plant, is valued for its exceptional nutritional profile, highlighting its monounsaturated fatty acid, omega-3, and antioxidant content. However, its application in the food industry faces challenges due to its susceptibility to oxidation and degradation during processing and storage [1]. To address this issue, encapsulation has emerged as an effective technique to protect and stabilize canola oil, thus preserving its nutritional and functional properties. This method allows for the formation of microcapsules containing small droplets of canola oil dispersed in a liquid or solid matrix, with potential applications in the food, pharmaceutical, and cosmetic industries. These microcapsules improve the stability, solubility, and bioavailability of canola oil in addition to protecting it from oxidation and other environmental factors. Canola oil has been shown to be effective in reducing low-density lipoprotein cholesterol (LDL-C) and apolipoprotein B levels in overweight and obese individuals, which is useful in preventing and reducing the risk of atherosclerosis [2]. Furthermore, its consumption has beneficial effects on LDL-C, total cholesterol, and the LDL-C–HDL-C ratio compared with olive oil, suggesting cardioprotective impacts [3].

Nanoemulsions have gained popularity in the food industry because they are manufactured from generally recognized as safe (GRAS) ingredients, offering advantages such as improved water dispersibility of the encapsulated oils, excellent physical and chemical stability, and high bioavailability of the lipid components. However, these unstable colloidal dispersions formed by two immiscible phases require the use of emulsifiers that facilitate emulsification, promote physical stability, and can influence oxidative stability, especially when there is the presence of pro-oxidant compounds such as transition metals that can interact with lipids in the core of the emulsion droplets [4,5]. Tween 80, a nonionic synthetic emulsifier with a high hydrophile–lipophile balance (HLB) value of 15.0, is widely used in the preparation of nano- and macro emulsions due to its excellent emulsifying properties, although its consumption is limited to an acceptable daily intake (ADI) of 25 mg/kg (172,840, U.S. FDA) [6–8]. On the other hand, soy lecithin, composed of a mixture of phospholipids, triglycerides, glycolipids, and sterols, is the most widely used natural emulsifier in the food industry. It is water soluble and forms a colloidal suspension that favors the stability of oil–water emulsions due to its hydrophilic and hydrophobic groups [9,10]. Both synthetic and natural emulsifiers play a crucial role in the stabilization of oil–water emulsions by adsorbing at the interface and reducing the interfacial tension, protecting the droplets against aggregation and allowing for the formation of stable and bioavailable systems.

Encapsulation is a process by which particles or droplets of bioactive compounds are surrounded by polymeric compounds, such as proteins and/or polysaccharides, to form small capsules containing the active compound of interest. This method is used to isolate, protect, release in a controlled manner, or mask the taste and odor of compounds such as oils [11,12]. Among the most commonly encapsulated ingredients are flavorings, lipids, enzymes, microorganisms, leavening agents, antioxidants, preservatives, colorants, minerals, essential oils, and vitamins. However, these compounds can be sensitive to degradation by light, temperature, or pH [13].

Sonication encapsulation is a process that uses ultrasonic energy to create microcapsules containing active components. Sonication applies high-frequency ultrasonic waves to a liquid mixture, generating cavitation bubbles that collapse rapidly. This phenomenon produces intense shear and turbulence forces useful for emulsifying and encapsulating oils in the form of microdroplets. Sonication is widely used in industry due to its high energy and efficiency [14]. This method offers several advantages, such as a fast and efficient process, the ability to encapsulate heat-sensitive compounds, and the production of microcapsules with a uniform particle size (between 6 and 15 μm) [15]. However, optimizing sonication conditions, such as intensity, power, frequency, and exposure time, is crucial to ensure the optimal manufacturing of high-quality microcapsules [16]. The development of matrices is essential in encapsulation, and protein substances can be used due to their excellent emulsification capacity, biocompatibility, and biological activity [17]. Polysaccharides can also be used, which allow for encapsulation using natural products, such as chitosan, carrageenan, whey proteins, lupine protein isolate [11], alginate [18], gums, pectins, and gelatins, or synthetic products, such as polyvinyl alcohol, polyethylene glycol, and polylactic-co-glycolic acid [13]. The interactions between proteins and polysaccharides improve the stability and sensitivity of the microcapsules [17]. Recent research has explored the encapsulation of active ingredients, such as hyssop extract (*Hyssopus officinalis* L.), obtained by extraction prior to ultrasound-assisted cold plasma treatment within a double emulsion stabilized with soy protein isolate (SPI) alone and combined with chia seed gum (CSG). In these studies, it was shown that canola oil containing encapsulated extract presented lower oxidative alterations than the non-encapsulated form [19]. In addition, tests have been performed on the antioxidant activity of aqueous extracts of *Bifurcaria bifurcata* (BBE) at different doses, compared to butylated hydroxytoluene (BHT) in canola oil, indicating that BBE may be a potential natural additive to improve the oxidative stability of canola oil [20]. In this context, the objective of this study was to evaluate the encapsulation of canola oil (*Brassica napus*) by sonication for the development of protein and starch systems in order to create stable nanoemulsified gels and determine the physical, colorimetric, and rheological stability of the encapsulated oil.

2. Materials and Methods

2.1. Materials

Oil-in-water (O/W) nanoemulsions were prepared using purified water from a reverse osmosis system (Vigaflow S.A., Santiago-Chile), canola oil (Belmont, Watt's SA, San Bernardo, Santiago, Chile), and two types of emulsifiers: soy lecithin (a natural emulsifier) (Metarin P-Cargill, Blumos SA, Santiago, Chile) and Tween 80 (a synthetic emulsifier) obtained from Sigma-Aldrich S.A. (St. Louis, CA, USA). In addition, protein isolate—WPI (Nutralys F85M, Lestron, France) and starch (Enterex Food, Santiago, Chile) were used for the preparation of the gels.

2.2. Stage 1

2.2.1. Preparation of Nanoemulsions

Oil-in-water (O/W) nanoemulsions were prepared using 5% canola oil and the emulsifiers soy lecithin (3% *w*/*w*) and Tween 80 at different concentrations (Table 1). The aqueous phase was prepared by dispersing Tween 80 in purified water using a magnetic stirrer (model Arex, Velp Scientifica, Usmate Velate, Italy) at 350 rpm for 10 min. Then, soy lecithin (3% *w*/*w*) was added to the aqueous phase and stirred for 40 min at 450 rpm until completely dispersed. Subsequently, canola oil was slowly added to the aqueous phase, while the mixture was homogenized at 12 000 rpm using a high-speed homogenizer (model IKA T25, Ultra Turrax, Germany) for 10 min. After adding all of the oil phase, the preemulsion was homogenized using an ultrasonic homogenizer (model VCX500, Sonics, Newtown, CT, USA) with a 13 mm diameter stainless steel ultrasound probe under conditions of 20 kHz, 90% amplitude, and between 5 and 20 min with work and rest intervals of 15 and 10 s, respectively. Finally, before measurements, the nanoemulsions were stored at 5 ± 1 °C for 24 h, with no destabilization phenomena observed under these conditions. The preparation procedure of the nanoemulsions was based on previous experiences in comparable rheological properties with similar products [4,21].

Table 1. Treatment formulation.

Treatment	Independent Variable			Dependent Variable
	Time (min)	Lecithin (g)	Tween 80 (g)	
1	12.5	2.45	0.3	Creaming index Particle size Polydispersion index
2	5.0	2.45	0.1	
3	12.5	0.40	0.5	
4	20.0	4.50	0.3	
5	20.0	2.45	0.5	
6	5.0	0.40	0.3	

Table 1. *Cont.*

Treatment	Independent Variable			Dependent Variable
	Time (min)	Lecithin (g)	Tween 80 (g)	
7	5.0	4.50	0.3	Creaming index Particle size Polydispersion index
8	20	0.40	0.3	
9	12.5	0.40	0.1	
10	12.5	2.45	0.3	
11	20.0	2.45	0.1	
12	12.5	2.45	0.3	
13	12.5	4.50	0.1	
14	5.0	2.45	0.5	
15	12.5	4.50	0.5	

2.2.2. Physical Properties for Determining Canola Oil Encapsulation

Particle Size Characterization and Polydispersity Index

The particle size and polydispersity index of different nanoemulsions were determined by dynamic light scattering (DLS) using a Zetasizer (NanoS90, Malvern Instruments, Malvern, UK). To perform the measurements, each nanoemulsion was diluted (50 µL of the nanoemulsion in 1 mL of milli-Q water) until a clear solution was obtained in order to reach a detectable concentration and avoid interference with the measurement. The particle size of the samples was described by the average particle size (PS), and the size distribution was described by the polydispersity index (PdI) [21]. The reported values are an average of 10 determinations and each sample was measured in triplicate ($n = 3$).

Physical Stability of Nanoemulsions

The instability of nanoemulsions during storage was characterized by the level of creaming, which is measured by the creaming index (H), where a high value of the creaming index is representative of a high degree of instability of the nanoemulsion. The creaming index was calculated according to Equation (1) as described by Petrovic et al. [22]:

$$Cream\ index\ (\%) = \frac{HS}{HE} \times 100\% \quad (1)$$

where HE is the total height of the nanoemulsion (mm) and HS is the height of the cream layer (mm), which was measured visually as a function of time. All measurements were performed in triplicate for each emulsion (n = 3).

2.2.3. Experimental Statistical Design

To evaluate the influence of sonication time, lecithin, and Tween 80, a Box–Behnken-type surface response design was used (Table 1). A total of 15 treatments were obtained (12 factorial points + 3 central points). From this design, a multiple response analysis was performed to find the optimal parameters for the physical stability of the nanoemulsion, which was applied in Stage 2.

2.3. *Stage 2*

2.3.1. Preparation of Nanoemulsion-Encapsulated Gels

For the preparation of gels with the nanoemulsion, the first step involved the formulation of the base nanoemulsion, which was optimized in the initial stage. This nanoemulsion consisted of 4 g of lecithin, 0.5 g of Tween 80, 5 g of canola oil, and 90.5 g of water sonicated for 10 min. Subsequently, different systems were prepared. The protein system, utilizing whey protein, was formulated at concentrations of 7.5, 10.0, 12.5, and 15.0%. In the case of the protein–starch mixture system, the protein concentration was kept constant at 3%, while the starch percentages varied over 1, 2, 3, and 4%. All components were mixed with 5% of the base nanoemulsion, and the remainder of the formulation was adjusted with water according to each treatment. Dispersion was performed using a propeller stirrer (F20100151, Velp Scientifica, Usmate Velate, Italy) at 450 rpm in a thermoregulated bath (Heating bath B-100, Buchi, Flawil, Switzerland) at 90 ± 1 °C for 30 min. The dispersions obtained were then cooled to room temperature (25 ± 1 °C) and stored at 4 ± 1 °C for 24 h until further characterization.

2.3.2. Colorimetric and Rheological Properties

Colorimetric Properties of Nanoemulsion-Based Gels

The optical properties of the nanoemulsion-encapsulated gels were determined using a colorimeter (ChromaMeter CR-410, Konica Minolta, Osaka, Japan), which was calibrated with a standard calibrated plate (L*: 93.46, a*: 0.42, b*: 4.08). A 20 g piece of each sample was placed in a Petri dish (60 mm internal diameter) to homogenize the surface and obtain the CIELab parameters L* (brightness), a* (redness–greenness), and b* (blueness–yellowness).

Flow Properties of Nanoemulsion-Based Gels

The flow behavior of all nanoemulsions encapsulated in a protein system or starch–protein mixture was characterized using a rotational rheometer (MCR 72, Anton Paar, Austria) equipped with a concentric cylinder's geometry (CC50, Anton Paar, Austria). Flow curves were determined by recording the shear stress (τ) vs. shear rate (γ) values of the samples from 1 to 100 s^{-1} and from 100 to 1 s^{-1} for 120 s. Measurements were performed at 25 ± 1 °C, controlled with a Peltier system [23]. The sample liquids were placed and allowed to stand for 10 min before measurement to recover their structure and reach the test temperature. In addition, the consistency (K) and flow (n) index were determined using the power law (Equation (2)) in order to compare the flow parameters of the samples.

$$\tau = K \cdot \gamma^n \quad (2)$$

Viscoelastic Properties of Nanoemulsion-Based Gels

Viscoelastic testing of the nanoemulsion-encapsulated gels was carried out at 25 ± 1 °C using a controlled stress rheometer (MCR 72, Anton-Paar, Austria) equipped with a plate geometry (PP 50, 50 mm diameter, 1 mm gap). After loading the sample into the rheometer, it was allowed to stand for 10 min to stabilize and reach the test temperature. Small-amplitude oscillation sweeps were performed to analyze the viscoelastic properties of the samples. First, the linear viscoelasticity zone (LVR) was determined, where strain sweeps between 0.01 and 100% were performed at 1 Hz. Then, a frequency sweep from 0.1 to 10 Hz was performed at 0.1% strain, which was selected as it lies within the LVR. Finally, oscillatory rheological parameters (G' and G'') were calculated at 1.17 Hz in order to compare the viscoelastic properties of the different nanoemulsion-encapsulated gels. Measurements were performed in triplicate (n = 3).

2.3.3. Experimental Statistical Design

A completely randomized design (CRD) was used to determine the differences between protein and starch systems (Table 2). Eight treatments were designed with different protein percentages and the protein–starch combination for the development of encapsulated gels with the base nanoemulsion.

Table 2. Formulation of nanoemulsion-based gel treatments.

Treatment	Independent Variable	Response Variables
Essay: Protein System (WPI)		
T1	7.5% WPI	
T2	10% WPI	
T3	12.5% WPI	
T4	15% WPI	Colorimetric parameters Rheological parameters
Essay: Protein–starch mixture system (ALM)		
T5	1% Starch + 3% WPI	
T6	2% Starch + 3% WPI	
T7	3% Starch + 3% WPI	
T8	4% Starch + 3% WPI	

2.4. Statistical Analysis

For the first stage, a three-factor ANOVA (analysis of variance) was performed using the Box–Behnken design and multi-response analysis of the creaming index, polydispersity index, and particle size. In the second stage, the colorimetric parameters and rheological properties were compared by applying the CRD through a one-way analysis and the Tukey test with a significance level of 95% using the R software. The experiments were performed in duplicate and each replicate was measured at least twice (minimum of four measurements). The results are reported as the average of all measurements and their corresponding standard deviation.

3. Results and Discussion

3.1. Stage 1—Physicochemical Characteristics of the Encapsulated Products

Table 3 presents the results obtained using the Box–Behnken design for the creaming index, particle size, and polydispersity index. Significant differences were found in the creaming index and particle size ($p > 0.05$), although it was not significant for the polydispersity index ($p > 0.05$). This design was used in several encapsulation studies [24–27]. These results are close to those obtained in peppermint oil emulsions stabilized with zein–lecithin complex nanoparticles with a particle size of 375 nm and a polydispersity index of 0.45 [28]. This difference is probably due to the type of encapsulation and the matrices used in the studies. Likewise, stable emulsions with canola protein isolates (*Brassica napus* L. var. *napus*), black caraway oil cake (*Nigella sativa*), and wheat bran (*Triticum aestivum*) produced lower and more consistent particle diameters [29]. Furthermore, nanoencapsulation of phenolic extract derived from the *Pimpinella affinis* plant, with *Salvia macrosiphon* gum and chitosan as coating agents, generated particle sizes between 75.8 and 144.2 nm and a variable polydispersity of 0.25 to 0.44 [30].

Table 4 shows the analysis of variance and the estimated regression coefficients corresponding to the physical stability parameters of the emulsions. For the creaming index, the model was significant ($p < 0.05$) for time (the quadratic term), lecithin concentration (the linear term), Tween 80 concentration (the linear and quadratic terms), and the time × Tween 80 interaction. The square term of time and Tween 80 concentration and the linear term of lecithin concentration significantly decreased ($p < 0.05$) the creaming index values, while the linear term of Tween 80 concentration and the time × Tween 80 interaction significantly increased it ($p < 0.05$). For the particle size parameter, the model was significant ($p < 0.05$). The lecithin concentration (linear and quadratic terms) and the Tween 80 concentration (linear term) significantly decreased ($p < 0.05$) this parameter. Regarding the polydispersity index, no significant effect ($p > 0.05$) of the factors and interactions was observed. Abbasi et al. [30] indicated that the differences between the emulsifying properties (surface activity, plasticity properties) of the coating materials can influence the results on the particle size of the emulsions.

Table 3. Physicochemical characteristics of the encapsulated product.

Treatments	Creaming Index (%)	Particle Size (nm)	Polydispersity Index
1	1.0	196.400	0.21525
2	4.5	258.325	0.26050
3	2.0	266.475	0.21825
4	1.0	161.400	0.21275
5	7.5	137.900	0.15775
6	5.5	280.800	0.21500
7	1.0	177.425	0.18350
8	5.5	273.900	0.21000
9	3.5	267.575	0.20900
10	1.5	174.275	0.15025
11	2.0	209.050	0.17800
12	1.0	172.325	0.16100
13	1.0	187.675	0.20200
14	4.5	143.575	0.15800
15	1.0	134.350	0.19025

Table 4. Regression coefficients and analysis of variance for the regression models corresponding to the physical stability parameters of the emulsions.

Factor	Creaming Index		Particle Size (nm)		Polydispersity Index	
	Coefficient	*p*-Value	Coefficient	*p*-Value	Coefficient	*p*-Value
Time	0.125	1.000	−19.469	0.176	−0.015	0.613
Time × Time	−2.416	0.004	−7.788	0.379	−0.007	0.745
Lecithin	−3.125	0.004	−106.975	0.008	−0.016	0.584
Lecithin × Lecithin	0.333	0.156	−34.594	0.038	−0.023	0.331
Tween 80	1.000	0.039	−60.081	0.024	−0.031	0.332
Tween 80 × Tween 80	−1.041	0.0201	1.575	0.842	−0.006	0.761
Time × Lecithin	0.000	1.000	−4.563	0.765	0.017	0.672
Time × Tween 80	2.750	0.011	21.800	0.245	0.041	0.359
Lecithin × Tween 80	0.750	0.121	−26.112	0.190	−0.011	0.792
R^2	0.845		0.932		0.556	
Lack of fit	0.0246		0.1956		0.567496	

Figure 1 shows the contour and response surface plots of the factors that influence the creaming index and droplet size. Figure 1a shows that the creaming index is influenced by intermediate sonication times working with high lecithin concentrations. This is corroborated by the significant effect of the time x lecithin interaction (Table 4). Similar behavior was observed in the creaming index with the sonication time and the Tween 80 concentration (Figure 1b), where intermediate times and low and intermediate concentrations of Tween 80 produce lower creaming parameters. Figure 1c presents the contour plot of soy lecithin and Tween 80, where higher concentrations of soy lecithin and low and intermediate values of Tween 80 reduce this parameter. Likewise, the R^2 value suggests that the model explains well the variation in the creaming index. These results can be contrasted with studies that evaluate the stability of emulsions and their relationship with stabilizing agents such as lecithin and surfactants (Tween 80) that affect stability by reducing the creaming index by decreasing the aggregation of droplet-sized particles, since this depends mainly on the type of thickener. Starch-based nanoemulsions showed lower particle size, polydispersity, and Z-potential values, close to zero [4].

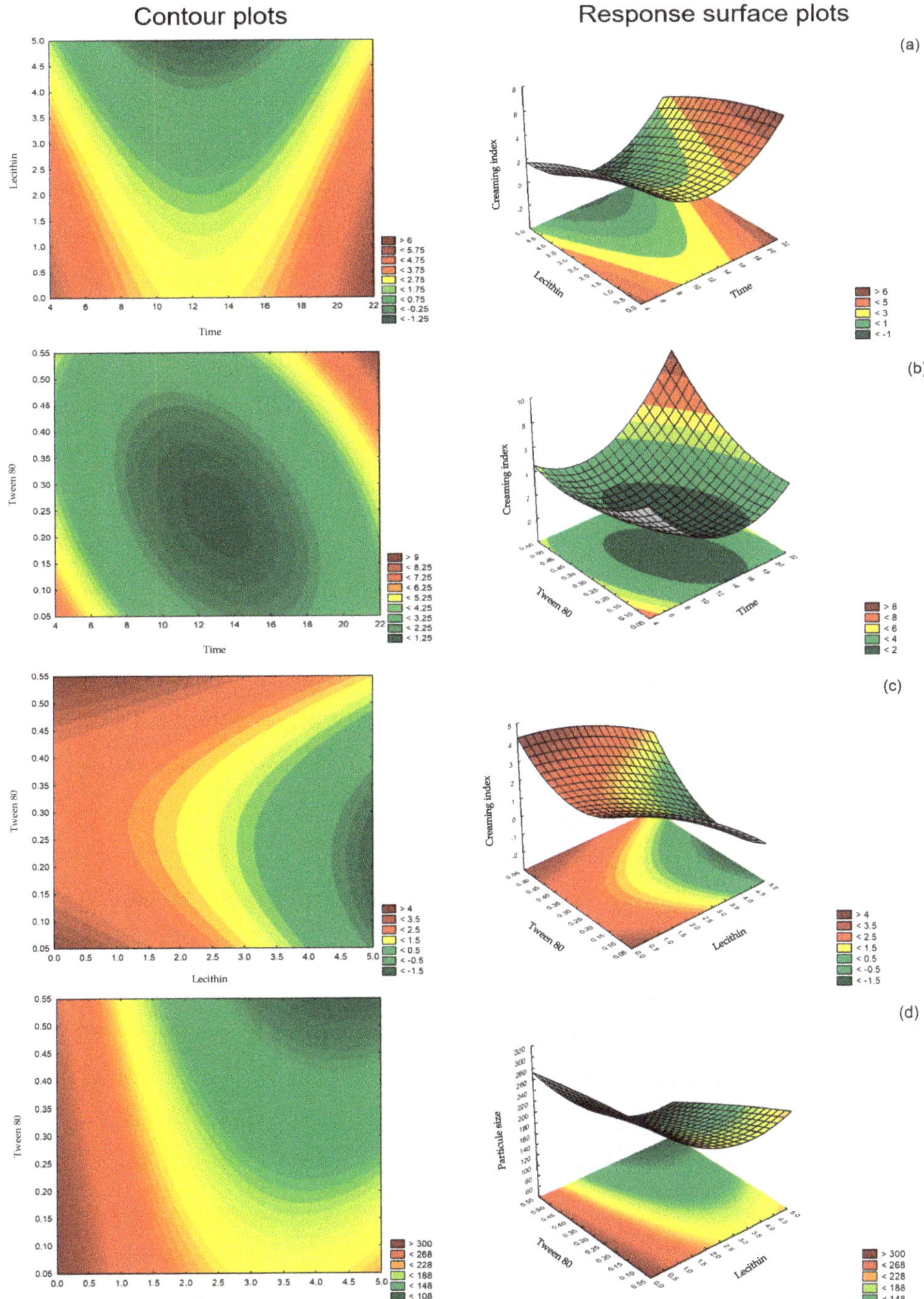

Figure 1. Contour plots and response surface for the creaming index. (**a**) Time vs. Lecithin, (**b**) Time vs. Tween 80, (**c**) Lecithin vs. Tween 80 and particle size, and (**d**) Lecithin vs. Tween 80.

Regarding particle size, Figure 1d presents the contour plot of soy lecithin and Tween 80, where higher soy lecithin concentrations and high Tween 80 values reduce this parameter. Furthermore, the negative coefficient of lecithin suggests that its addition significantly reduces the droplet size, which is consistent with its effect as an emulsifier. On the other hand, the R^2 shows that the model adequately describes the particle size (Table 4). These findings agree with studies on emulsions in which both lecithin and Tween 80 were used to stabilize and reduce the particle size. In works carried out with Tween 80 at a surfactant-to-oil ratio ≥ 1 at high stirring speeds (800 rpm), nanoemulsions with small particle diameters (d < 200 nm) were generated, being systems relatively stable to particle growth at room temperature (<10% of diameter after 1 month of storage) but unstable to heating (T > 80 °C) [7]. Also, extensive aggregation between oil particles gives rise to unstable emulsions due to the interactions produced [31].

Finally, as indicated, the polydispersity index is not significantly affected by any of the main factors or their interactions, since all p values are higher than 0.05. Furthermore, the lower R^2 also suggests that the model does not explain the variability in the polydispersity index as well. This could be interpreted as the formulations used providing sufficient stability, minimizing the differences in polydispersity.

In Figure 2, the graph generated by the desirability function is presented, which is one of the most used methods for optimizing the effect of multiple factors on a response variable. This method was developed by Harrington (1965) and modified by Derringer and Suich (1980), being applied in various studies on the optimization of encapsulation processes [32,33]. The desirability function was used to optimize the experimental design factors (lecithin, Tween 80 concentration, and sonication time) in order to obtain the desired values of the evaluated parameters. Figure 2 shows the graph of the analysis with the desirability function used to obtain the minimum value of the response variables for the development of the nanoemulsion. The conditions of the independent variables were 4.0 g of lecithin, 0.5 of Tween 80, and 10 min of sonication time, where the following predicted values were obtained for the response variables: a creaming index of 1.165%, a particle size of 135.081 nm, and a sonication time of 10 min.

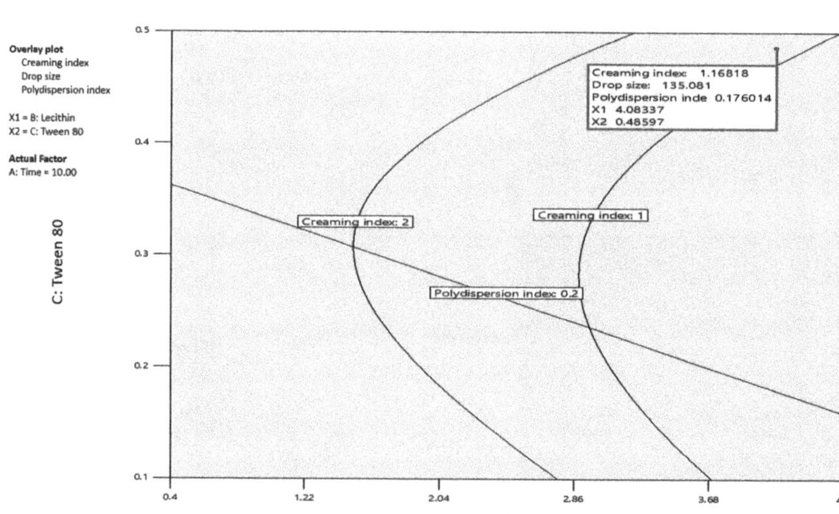

Figure 2. Multiple response analysis to find the optimal parameters for nanoemulsion development.

3.2. Stage 2—Colorimetric and Rheological Characteristics of the Gels with the Nanoemulsion

3.2.1. Colorimetric Parameters

The variance analysis of the emulsion's chromatic parameters is presented in Figure 3. We observed significant differences ($p < 0.05$) for the different treatments evaluated, mainly the concentrations of WPI alone and mixtures of starch with WPI. Considering the CIELab coordinates, the color of a gel with the nanoemulsion is generated when a beam of light hits the oil droplets and disperses the light. This depends largely on the oil concentration [34]. However, it is not the only factor that influences its appearance. The other components of the emulsion are also influential.

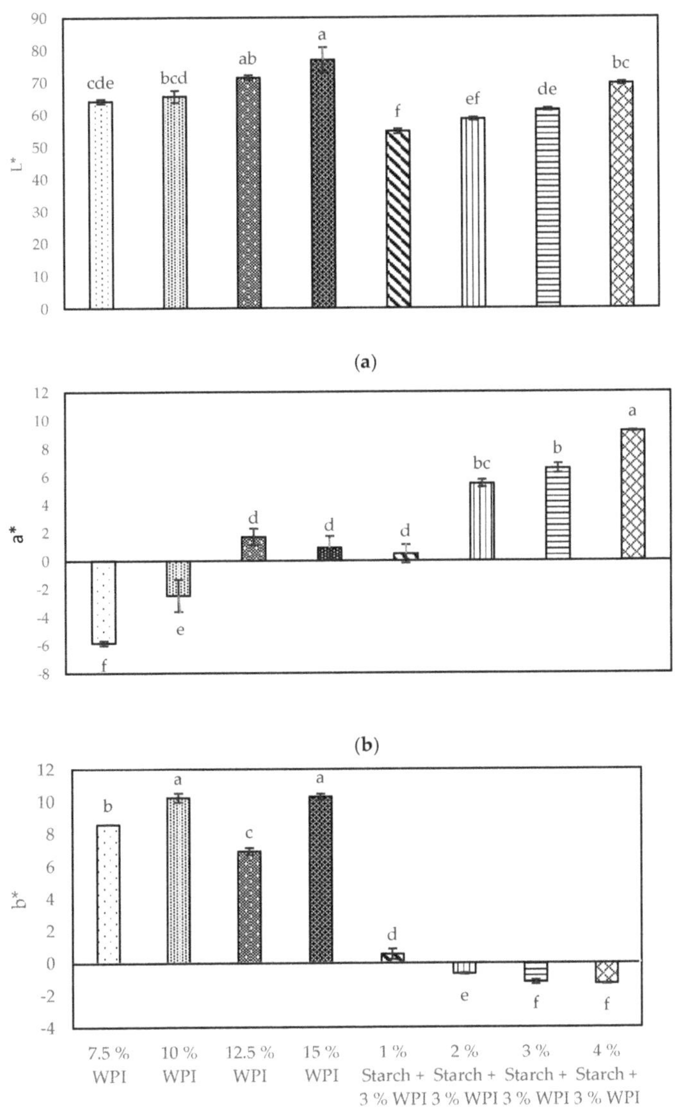

Figure 3. Colorimetric parameters of nanoemulsion-encapsulated gels. (**a**) L*, (**b**) a*, (**c**) b* (Note: a,b,c,d,e,f Different letters indicate significant differences).

The luminosity (L*) of the gel with the nanoemulsion was significantly higher ($p < 0.05$) due to the effect of the 12.5 and 15.0% WPI. In turn, the lowest values for this parameter were obtained due to the effect of 1% starch + 3% WPI and due to the effect of 2% starch + 3% WPI. In similar studies, the L* was influenced by the size of the drop, giving higher luminosity values when the diameter of the drop increases, generating greater light dispersion. On the other hand, smaller particles scatter less light [35].

The emulsion parameter a* decreased significantly ($p < 0.05$) due to the effect of the 7.5% WPI, while the highest value was obtained with 4% starch + 3% WPI. Therefore, the qualitative color obtained for the nanoemulsions was yellow. The a* values shown in Figure 3 ranged from −6 to 10. Because this variable varied linearly with increasing concentrations of WPI and starch, the values tended to be more positive, representing a shift toward red [36].

The parameter b* decreased significantly ($p < 0.05$) due to the effect of 3 and 4% starch + 3% WPI, while 10 and 15% WPI increased this parameter significantly ($p < 0.05$). The b* value, due to the qualitatively yellow color of the nanoemulsions, shifted toward blue (minimum b* values), fluctuating from −10 to 2 when the starch concentrations were high and the WPI concentrations were low [36].

3.2.2. Rheological Parameters

The results obtained on the rheological properties are presented in Figure 4. Specific behaviors were identified in the viscosity and shear stress of the samples analyzed at different shear rates and different concentrations of protein system gels and starch mixtures with WPI.

Figure 4a shows the relationship between shear rate and shear stress for different concentrations of proteinaceous gels (WPI) and starch–WPI blends (ALM). For all concentrations, the shear stress increased with the shear rate, being similar to the reduction in rheological properties in soy protein isolate and corn starch gels [37]. This indicates that WPI and ALM experience an increase in the stress required to deform as the shear rate increases. The differences in shear stress show that the 15% WPI sample and the 4% ALM sample have the highest shear stress values throughout the shear rate range. This suggests that both systems, at these concentrations, require higher stress to be deformed; thus, the gel strength is almost opposite to the viscosity order [38]. The lower concentrations of WPI (7.5, 10, and 12.5) show lower shear stresses than 15% WPI, but with a similar trend of increasing with the shear rate. In the case of ALM, the concentrations of 1, 2, and 3% show lower shear stress values compared with 4% ALM, but they also increase steadily with shear rate. The higher concentrations of both substances (WPI and ALM) have higher shear stress values, indicating that the resistance to deformation increases with concentration. WPI appears to be more sensitive to concentration than ALM in terms of shear stress, as the WPI samples exhibited a more pronounced increase in shear stress at higher concentrations.

Figure 4b shows the relationship between shear rate and viscosity for different concentrations of WPI and ALM-based emulsions. The shear behavior shows that, for each concentration, the viscosity decreases as the shear rate increases. This suggests that both WPI and ALM exhibit shear thinning behavior, meaning that they become less viscous when subjected to higher shear rates. The 15% WPI solution had the highest initial viscosity, starting above 14 000 mPa s and showing a steep drop as the shear rate increased. The lower WPI concentrations (7.5, 10, and 12.5%) showed decreasing initial viscosities with a similar decreasing trend as the shear rate increased. The starch solutions (1, 2, and 3% ALM) had lower initial viscosities than the WPI solutions at corresponding concentrations. However, among the WPI starch samples, the 4% ALM one had the highest viscosity, similar to the 15% WPI gel, while the 1% ALM gel had the lowest viscosity. Therefore, higher concentrations of both WPI and ALM generally result in higher initial viscosities. WPI appears to have a more concentration-dependent viscosity response than ALM. It is evident that both WPI and ALM exhibit shear-thinning behavior, with WPI showing significantly higher viscosities at elevated concentrations compared with ALM.

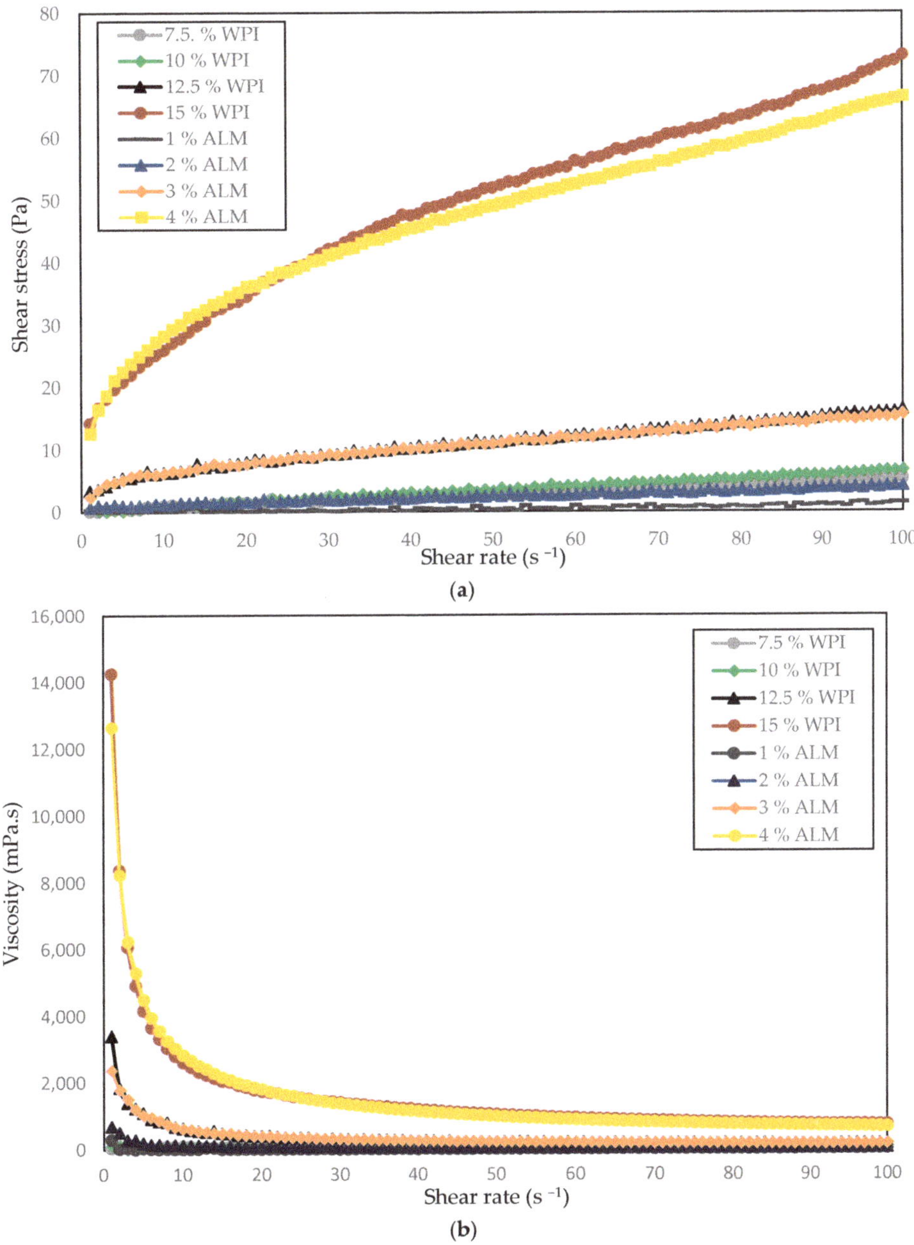

Figure 4. Flow curves of gels of protein systems and starch mixtures. (**a**) Shear stress vs. shear rate and (**b**) viscosity vs. shear rate.

Figure 5 shows the storage modulus G′ and the loss modulus G″ (in Pa) of different concentrations of WPI and ALM. The viscoelastic properties were determined within the linear viscoelasticity region (LVR) [38,39]. This allowed us to obtain more information on the structure of the gels at different concentrations of WPI and storage periods with WPI, managing to generate protein systems (WPI) and systems based on mixtures of starch and protein (ALM). The values of G′–G″ indicate the elasticity and viscosity of the material. The viscoelastic behavior indicated that G′ is significantly higher than G″. This indicates a predominance of elastic behavior in the network, which is desirable for structural applications. The viscoelasticity values were higher at higher concentrations (15% WPI and 4% ALM). Furthermore, the distance between G′ and G″ provides additional information about the network quality: a smaller distance suggests a more cohesive and stable structure [40]. The values of G′–G″ increased with frequency in all cases. As the WPI concentration increased (7.5 to 15%), the G′–G″ values also increased, indicating higher stiffness or elasticity at these higher concentrations. For ALM, a similar pattern was observed, where higher concentrations (e.g., 4% ALM) showed higher values in G′–G″, suggesting similar behavior in terms of elasticity and energy storage capacity. This behavior of both the elastic (G′) and viscous (G″) moduli of the gels increased linearly with increasing frequency, revealing a strong frequency dependence [41]. These results indicate that both systems (WPI and ALM) have the potential to form low-strength gels incorporating nanoemulsions, which require minimal force to bite and chew, easily disintegrating during processing in the mouth. Therefore, these gels are suitable for the development of food products with excellent textural properties.

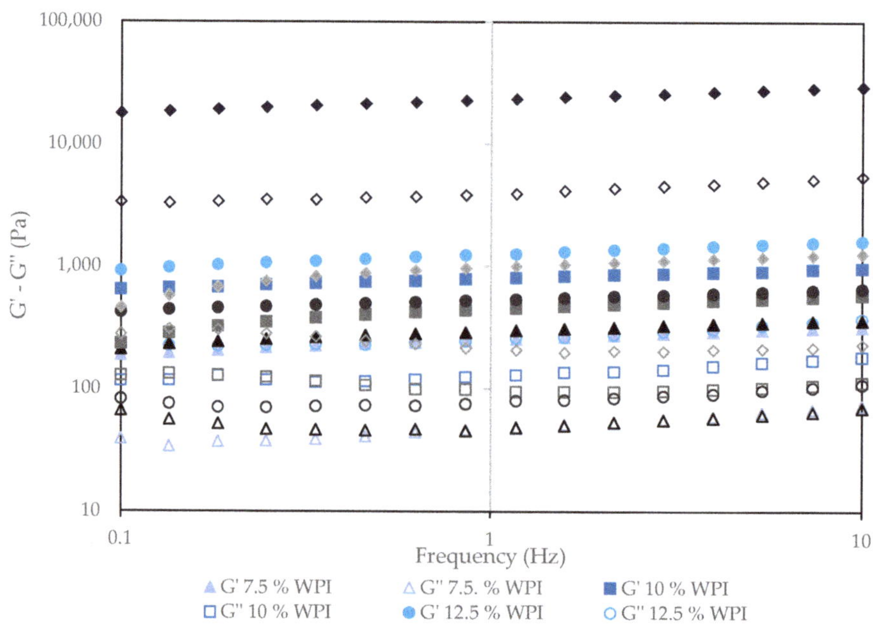

Figure 5. Frequency sweep from 0.1 to 10 Hz of gels of protein systems and starch mixtures.

In Table 5, the values of the rheological parameters of the protein (WPI) and starch–protein mixture (ALM) systems are shown. For the flow sweep data, the consistency index (K) and apparent viscosity were found to have significant differences ($p < 0.05$) between the treatments. The K values increased with the WPI concentration, from 0.086 at 7.5% to 11.069

at 15.0%. In the samples with ALM, it also increased with concentration, from 0.019 at 1.0% to 12.422 at 4.0%, suggesting that ALM has a greater capacity to form a more resistant structure at high concentrations. Regarding the flow index (n), it was found to be different between all treatments ($p < 0.05$). The n values for the WPI decreased with increasing concentration, going from 0.897 to 0.393, indicating that it becomes more pseudoplastic (stress-thinning flow). For ALM, it also decreased (from 0.826 at 1% to 0.355 at 4%), showing similar behavior. All treatments had a high R^2 (coefficient of determination) (close to 1), indicating a good fit of the data to the power law.

Table 5. Values of rheological parameters using the power law and constant frequency sweep.

Treatment	Flow Sweep—Power Law			Oscillatory Sweep—Frequency 1.17 Hz	
	K	n	R^2	G'	G''
T1: 7.5% WPI	0.086 ± 0.0002 [e]	0.897 ± 0.001 [a]	0.971 ± 0.0002	257.950 ± 3.320 [c]	47.793 ± 0.576 [g]
T2: 10% WPI	0.085 ± 0.0004 [e]	0.851 ± 0.0016 [b]	0.918 ± 0.0006	807.110 ± 4.790 [bc]	129.890 ± 0.997 [d]
T3: 12.5% WPI	2.781 ± 0.0016 [c]	0.364 ± 0.0015 [g]	0.981 ± 0.0006	1258.400 ± 2.400 [b]	262.080 ± 1.030 [b]
T4: 15% WPI	11.069 ± 0.0969 [b]	0.393 ± 0.0019 [e]	0.988 ± 0.0011	22,987.000 ± 685.00 [a]	3922.900 ± 9.05 [a]
T5: 1% Starch + 3% WPI	0.019 ± 0.0001 [e]	0.826 ± 0.0008 [c]	0.802 ± 0.0003	303.610 ± 1.970 [bc]	48.627 ± 0.528 [g]
T6: 2% Starch + 3% WPI	0.478 ± 0.0006 [d]	0.439 ± 0.0007 [d]	0.929 ± 0.0004	458.910 ± 1.820 [bc]	94.904 ± 0.560 [e]
T7: 3% Starch + 3% WPI	2.639 ± 0.0016 [c]	0.374 ± 0.0006 [f]	0.987 ± 0.0003	528.060 ± 2.740 [bc]	78.503 ± 0.703 [f]
T8: 4% Starch + 3% WPI	12.422 ± 0.0318 [a]	0.355 ± 0.0001 [h]	0.997 ± 0.0003	1000.900 ± 1.520 [bc]	208.110 ± 0.559 [c]

[a–h] Different letters indicate significant differences.

For the frequency sweep, G' represents the storage modulus, which is the elasticity of the material or the ability to store energy. In the case of WPI, the G' increased remarkably with concentration, from 257.95 at 7.5% to 22,987 at 15%, showing a significant increase in elasticity. For ALM, the G' also increased with concentration, although the values were lower than those of the WPI at high concentrations. The maximum value in ALM was 1000.9 at 4%. In addition, G' represents the loss modulus, which represents the viscosity of the material or its ability to dissipate energy. In the WPI, the G'' also increased with concentration, from 47.79 at 7.5% to 3922.9 at 15%. In the ALM, the G'' also increased with concentration, although with lower values than in the WPI, reaching a maximum of 208.11 at 4.0%. On the other hand, the elastic modulus (G') was generally higher in the WPI than in the ALM. This indicates that WPI exhibits greater elasticity and stiffness compared with ALM, especially at high concentrations. This behavior suggests that WPI has a superior ability to recover its original shape after deformation, contributing to a more stable structure in applications where mechanical strength is required. Likewise, the viscosity (G'') indicated that the G'' values are also higher in WPI, suggesting that WPI has a higher energy dissipation capacity than ALM. WPI seems to form a stronger viscoelastic network compared with ALM, especially at higher concentrations. At all concentrations, the G' values exceeded the G'' values, indicating elastic solid behavior with stable and well-structured networks [41]. Both the G' values and G'' values increased with frequency, meaning that the gel network structure may be formed through non-covalent interactions [40].

4. Conclusions

Nanoemulsions present physicochemical characteristics that are significantly influenced by the concentration of lecithin and Tween 80, as well as by the sonication time, which translates into a reduced particle size and an optimized creaming index through a Box–Behnken experimental design. A higher concentration of lecithin and Tween 80 reduces the droplet size, which is consistent with its function as an effective and stable long-term emulsifier. The desirability function used to optimize the experimental conditions allowed us to identify formulations that maximize stability, highlighting the potential of these nanoemulsions to improve the functional quality in food products. Gels with nanoemulsions made from whey protein isolate (WPI) and starch mixtures with WPI (ALM) present colorimetric and rheological characteristics that are significantly influenced by the concentration of lecithin and Tween 80, as well as by the sonication time. An increase in WPI concentration increases the lightness and affects the colorimetric parameters, resulting in emulsions with a more pronounced yellow color. Furthermore, the rheological properties indicate that both WPI and starch exhibit shear thinning behavior, being more pronounced at high concentrations, suggesting that these gels are suitable for food applications where a smooth texture and rapid disintegration are required.

Author Contributions: Conceptualization, methodology, and writing—original draft preparation, C.A. and R.J.S.-P.; data curation, C.A., N.R. and T.A.R.-A.; resources, C.E.Ñ.-Q. and N.C.J.-G.; statistical analysis, R.J.S.-P. and A.E.-S.; formal analysis, R.J.S.-P. and C.A.; writing—review and editing and visualization, R.J.S.-P., N.C.J.-G., T.A.R.-A. and C.A.; supervision, validation, project administration, and funding acquisition, R.J.S.-P. All authors have read and agreed to the published version of the manuscript.

Funding: This research was funded by Advanced Studies of the National Council for Science, Technology, and Technological Innovation (PROCIENCIA-CONCYTEC) through contract No. PE501082112-2023-PROCIENCIA for Applied Research Projects.

Institutional Review Board Statement: Not applicable.

Informed Consent Statement: Not applicable.

Data Availability Statement: The original contributions presented in the study are included in the article. Further inquiries can be directed to the corresponding authors.

Acknowledgments: This research was developed at the facilities of the National University of Barranca and the University of Santiago de Chile with funding from the National Program for Scientific Research.

Conflicts of Interest: The authors declare no conflicts of interest.

References

1. Enns, J.E.; Zahradka, P.; Guzman, R.P.; Baldwin, A.; Foot, B.; Taylor, C.G. Randomized controlled trial to evaluate the effect of canola oil on blood vessel function in peripheral arterial disease: Rationale and design of the Canola-PAD Study. *Open Access J. Clin. Trials* **2014**, *6*, 117–125.
2. Yang, J.-M.; Long, Y.; Ye, H.; Wu, Y.-L.; Zhu, Q.; Zhang, J.-H.; Huang, H.; Zhong, Y.-B.; Luo, Y.; Wang, M.-Y. Effects of rapeseed oil on body composition and glucolipid metabolism in people with obesity and overweight: A systematic review and meta-analysis. *Eur. J. Clin. Nutr.* **2024**, *78*, 6–18. [CrossRef]
3. Pourrajab, B.; Sharifi-Zahabi, E.; Soltani, S.; Shahinfar, H.; Shidfar, F. Comparison of canola oil and olive oil consumption on the serum lipid profile in adults: A systematic review and meta-analysis of randomized controlled trials. *Crit. Rev. Food Sci. Nutr.* **2023**, *63*, 12270–12284. [CrossRef]
4. Arancibia, C.; Miranda, M.; Matiacevich, S.; Troncoso, E. Physical properties and lipid bioavailability of nanoemulsion-based matrices with different thickening agents. *Food Hydrocoll.* **2017**, *73*, 243–254. [CrossRef]
5. Alipour, E.; Halverson, D.; McWhirter, S.; Walker, G.C. Phospholipid Bilayers: Stability and Encapsulation of Nanoparticles. *Annu. Rev. Phys. Chem.* **2017**, *68*, 261–283. [CrossRef] [PubMed]
6. McClements, D.J. Nanoemulsion-based oral delivery systems for lipophilic bioactive components: Nutraceuticals and pharmaceuticals. *Ther. Deliv.* **2013**, *4*, 841–857. [CrossRef] [PubMed]
7. Guttoff, M.; Saberi, A.H.; Mcclements, D.J. Formation of vitamin D nanoemulsion-based delivery systems by spontaneous emulsification: Factors affecting particle size and stability. *Food Chem.* **2015**, *171*, 117–122. [CrossRef] [PubMed]
8. Raikos, V. Encapsulation of vitamin E in edible orange oil-in-water emulsion beverages: Influence of heating temperature on physicochemical stability during chilled storage. *Food Hydrocoll.* **2017**, *72*, 155–162. [CrossRef]
9. Mezdour, S.; Desplanques, S.; Relkin, P. Effects of residual phospholipids on surface properties of a soft-refined sunflower oil: Application to stabilization of sauce-types' emulsions. *Food Hydrocoll.* **2011**, *25*, 613–619. [CrossRef]

10. Ozturk, B.; McClements, D. Progress in natural emulsifiers for utilization in food emulsions. *Curr. Opin. Food Sci.* **2016**, *7*, 1–6. [CrossRef]
11. Morales, E.; Burgos-Díaz, C.; Zúñiga, R.N.; Jorkowski, J.; Quilaqueo, M.; Rubilar, M. Influence of O/W emulsion interfacial ionic membranes on the encapsulation efficiency and storage stability of powder microencapsulated astaxanthin. *Food Bioprod. Process.* **2021**, *126*, 143–154. [CrossRef]
12. CGNA (Director). Microencapsulación: Una Tecnología que Potencia la Calidad y la Innovación de los Alimentos. Available online: https://www.youtube.com/watch?v=iYWeBJOWX10 (accessed on 10 August 2024).
13. Barbosa-Nuñez, J.A.; Espinosa-Andrews, H.; Cardona AA, V.; Haro-González, J.N. Polymer-based encapsulation in food products: A comprehensive review of applications and advancements. *J. Future Foods* **2025**, *5*, 36–49. [CrossRef]
14. Deng, R.-X.; Zheng, Y.-Y.; Liu, D.-J.; Liu, J.-Y.; Zhang, M.-N.; Xi, G.-Y.; Song, L.-L.; Liu, P. The effect of ultrasonic power on the physicochemical properties and antioxidant activities of frosted figs pectin. *Ultrason. Sonochem.* **2024**, *106*, 106883. [CrossRef]
15. Ho, J.; Wang, H.; Forde, G.M. Process considerations related to the microencapsulation of plasmid DNA via ultrasonic atomization. *Biotechnol. Bioeng.* **2008**, *101*, 172–181. [CrossRef]
16. Qu, W.; Feng, Y.; Xiong, T.; Qayum, A.; Ma, H. Preparation, structural and functional characterization of corn peptide-chelated calcium microcapsules using synchronous dual frequency ultrasound. *Ultrason. Sonochem.* **2024**, *102*, 106732. [CrossRef]
17. Ma, D.; Yang, B.; Zhao, J.; Yuan, D.; Li, Q. Advances in protein-based microcapsules and their applications: A review. *Int. J. Biol. Macromol.* **2024**, *263*, 129742. [CrossRef]
18. Piornos, J.A.; Burgos-Díaz, C.; Morales, E.; Rubilar, M.; Acevedo, F. Highly efficient encapsulation of linseed oil into alginate/lupin protein beads: Optimization of the emulsion formulation. *Food Hydrocoll.* **2017**, *63*, 139–148. [CrossRef]
19. Ahmadian, S.; Kenari, R.E.; Amiri, Z.R.; Sohbatzadeh, F.; Khodaparast MH, H. Fabrication of double nano-emulsions loaded with hyssop (*Hyssopus officinalis* L.) extract stabilized with soy protein isolate alone and combined with chia seed gum in controlling the oxidative stability of canola oil. *Food Chem.* **2024**, *430*, 137093. [CrossRef]
20. Agregán, R.; Lorenzo, J.M.; Munekata PE, S.; Dominguez, R.; Carballo, J.; Franco, D. Assessment of the antioxidant activity of Bifurcaria bifurcata aqueous extract on canola oil. Effect of extract concentration on the oxidation stability and volatile compound generation during oil storage. *Food Res. Int.* **2017**, *99*, 1095–1102. [CrossRef] [PubMed]
21. Arancibia, C.; Navarro-Lisboa, R.; Zúniga, R.N.; Matiacevich, S. Application of CMC as thickener on nanoemulsions based on olive oil: Physical properties and stability. *Int. J. Polym. Sci.* **2016**, *2016*, 6280581. [CrossRef]
22. Petrovic, L.B.; Sovilj, V.J.; Katona, J.M.; Milanovic, J.L. Influence of polymer-surfactant interactions on o/w emulsion properties and microcapsule formation. *J. Colloid Interface Sci.* **2010**, *342*, 333–339. [CrossRef] [PubMed]
23. Wang, Q.; Zhang, F.; Wang, S.; Chen, W.; Li, X.; Hao, J.; Alouk, I.; Wang, Y.; Xu, D.; Sun, B. The fabrication, microstructure, rheological properties and interactions of soft solid oleogels of hazelnut oil body. *Food Hydrocoll.* **2025**, *159*, 110711. [CrossRef]
24. Feczkó, T.; Tóth, J.; Dósa, G.; Gyenis, J. Optimization of protein encapsulation in PLGA nanoparticles. *Chem. Eng. Process. Process Intensif.* **2011**, *50*, 757–765. [CrossRef]
25. Bouriche, S.; Cózar-Bernal, M.J.; Rezgui, F.; Álvarez, A.M.R.; González-Rodríguez, M.L. Optimization of preparation method by W/O/W emulsion for entrapping metformin hydrochloride into poly (lactic acid) microparticles using Box-Behnken design. *J. Drug Deliv. Sci. Technol.* **2019**, *51*, 419–429. [CrossRef]
26. Essifi, K.; Lakrat, M.; Berraaouan, D.; Fauconnier, M.-L.; El Bachiri, A.; Tahani, A. Optimization of gallic acid encapsulation in calcium alginate microbeads using Box-Behnken Experimental Design. *Polym. Bull.* **2021**, *78*, 5789–5814. [CrossRef]
27. Sharaf, N.S.; Shetta, A.; Elhalawani, J.E.; Mamdouh, W. Application of Box-Behnken design for the formulation and optimization of coffee and PLGA nanoparticles and detection of enhanced antioxidant and anticancer activities. *Polymers* **2022**, *14*, 144. [CrossRef]
28. Xie, H.; Ni, F.; Gao, J.; Liu, C.; Shi, J.; Ren, G.; Tian, S.; Lei, Q.; Fang, W. Preparation of zein-lecithin-EGCG complex nanoparticles stabilized peppermint oil emulsions: Physicochemical properties, stability and intelligent sensory analysis. *Food Chem.* **2022**, *383*, 132453. [CrossRef] [PubMed]
29. Aktaş, H.; Custodio-Mendoza, J.; Moczkowska-Wyrwisz, M.; Szpicer, A.; Kurek, M.A. The role of canola, black caraway, and wheat bran protein isolates in anthocyanin microencapsulation via double emulsions. *Ind. Crops Prod.* **2024**, *222*, 119644. [CrossRef]
30. Abbasi, H.; Tavakoli, J.; Zare, F.; Salmanpour, M. Improving the efficacy of phenolic extract from *Pimpinella affinis* in edible oils through nanoencapsulation: Utilizing chitosan and *Salvia macrosiphon* gum as coating agents. *Food Sci. Nutr.* **2024**, *12*, 5463–5472. [CrossRef]
31. Paximada, P.; Koutinas, A.A.; Scholten, E.; Mandala, I.G. Effect of bacterial cellulose addition on physical properties of WPI emulsions. Comparison with common thickeners. *Food Hydrocoll.* **2016**, *54*, 245–254. [CrossRef]
32. Li, X.; Wang, L.; Wang, B. Optimization of encapsulation efficiency and average particle size of *Hohenbuehelia serotina* polysaccharides nanoemulsions using response surface methodology. *Food Chem.* **2017**, *229*, 479–486. [CrossRef] [PubMed]

33. Pashazadeh, H.; Zannou, O.; Ghellam, M.; Koca, I.; Galanakis, C.M.; Aldawoud, T.M. Optimization and encapsulation of phenolic compounds extracted from corn waste by freeze-drying, spray drying and microwave drying using maltodextrin. *Foods* **2021**, *10*, 1396. [CrossRef]
34. Salvia-Trujillo, L.; Rojas-Graü, M.A.; Soliva-Fortuny, R.; Martín-Belloso, O. Use of antimicrobial nanoemulsions as edible coatings: Impact on safety and quality attributes of fresh-cut Fuji apples. *Postharvest Biol. Technol.* **2015**, *105*, 8–16. [CrossRef]
35. McClements, D.J. Colloidal basis of emulsion color. *Curr. Opin. Colloid Interface Sci.* **2002**, *7*, 451–455. [CrossRef]
36. Ricaurte, L.; de Jesús Perea-Flores, M.; Martinez, A.; Quintanilla-Carvajal, M.X. Production of high-oleic palm oil nanoemulsions by high-shear homogenization (microfluidization). *Innov. Food Sci. Emerg. Technol.* **2016**, *35*, 75–85. [CrossRef]
37. Pi, X.; Zhu, L.; Xiang, M.; Zhao, S.; Cheng, Z.; Qiao, D.; Zhang, B. Insight of soy protein isolate to decrease the gel properties corn starch based binary system: Rheological and structural investigation. *Food Hydrocoll.* **2025**, *160*, 110750. [CrossRef]
38. Fu, J.; Cai, X.; Yang, Y.; Xie, H.; Duan, Q.; Liu, H.; Yu, L. Application of various polysaccharide gums to improve gelation and rheological properties of hydroxypropyl starch hydrocolloids. *Food Hydrocoll.* **2024**, *154*, 110043. [CrossRef]
39. Liang, X.; Ma, C.; Yan, X.; Zeng, H.; McClements, D.J.; Liu, X.; Liu, F. Structure, rheology and functionality of whey protein emulsion gels: Effects of double cross-linking with transglutaminase and calcium ions. *Food Hydrocoll.* **2020**, *102*, 105569. [CrossRef]
40. Farooq, S.; Ahmad, M.I.; Zhang, Y.; Chen, M.; Zhang, H. Preparation, characterization and digestive mechanism of plant-derived oil bodies-based oleogels structured by chitosan and vanillin. *Food Hydrocoll.* **2023**, *136*, 108247. [CrossRef]
41. Mert, B.; Vilgis, T.A. Hydrocolloid coated oleosomes for development of oleogels. *Food Hydrocoll.* **2021**, *119*, 106832. [CrossRef]

colloids and interfaces

MDPI

Article

Exploring the Safety of the Sustainable Toxicity Testing in Zebrafish and Brine Shrimp Using Nanoemulsions Formulated from Fish Byproducts and Lemon Oil

Amira Ayman Hendawy [1], Amal A. M. Elgharbawy [1,2,*], Najihah Mohd Noor [2], Nurhidayu Al-Saari [1,*], Nor Azrini Nadiha Azmi [3] and Hamzah Mohd Salleh [4,*]

1 International Institute for Halal Research and Training (INHART), International Islamic University Malaysia, Gombak 53100, Kuala Lumpur, Malaysia
2 Department of Chemical Engineering and Sustainability, Kulliyyah of Engineering, International Islamic University Malaysia, Gombak 50728, Kuala Lumpur, Malaysia; najihahmnoor@gmail.com
3 Halal Products Research Institute, Universiti Putra Malaysia, Serdang 43400, Selangor, Malaysia; azriniazmi@gmail.com
4 Halalan Thayyiban Research Centre, Sultan Sharif Ali Islamic University, Bandar Seri Begawan BE1310, Brunei
* Correspondence: amalgh@iium.edu.my (A.A.M.E.); hidayusaari@iium.edu.my (N.A.-S.); hamzah.salleh@unissa.edu.bn (H.M.S.)

Abstract: Nanoemulsions, characterized by their nanosized particles ranging from 20 to 200 nm, are effective carriers for drug molecules. Our novel oil-in-water nanoemulsion, NE-FLO™, formulated from lemon and fish byproduct oils, demonstrates promising antioxidant and anti-inflammatory activities, with initial studies indicating nontoxicity to normal skin cells. This study investigated the safety of NE-FLO™ using brine shrimp (*Artemia salina*) and zebrafish (*Danio rerio*) models, focusing on concentration-dependent effects and LC_{50} values. At lower concentrations (0.1 mg·L^{-1}, 0.01 mg·L^{-1}, and 0.001 mg·L^{-1}), NE-FLO™ showed minimal toxicity without adverse effects. However, at 1 mg·L^{-1}, reduced survival rates indicate potential toxicity. Specifically, this concentration also induces altered swimming behaviors in zebrafish. LC_{50} values are 8.7474 mg·L^{-1} for brine shrimp and 0.316 mg·L^{-1} for adult zebrafish. These results underscore the necessity for further detailed investigations into NE-FLO™, balancing its therapeutic benefits with potential toxicity risks. This study emphasizes the importance of optimizing nanoemulsion formulations from fish oil and conducting comprehensive safety assessments to meet regulatory standards.

Keywords: nanoemulsion; zebrafish; toxicity; sustainability; fish oil; safety

1. Introduction

Nanoemulsion technology has garnered significant attention for its diverse applications and potential in promoting sustainable practices. Fish oil, renowned for its health benefits, faces challenges due to its hydrophobic nature and poor bioavailability. To overcome these limitations, fish oil nanoemulsions—nanoscale droplets of fish oil dispersed in water—have been developed. These nanoemulsions have shown promise as effective delivery systems for omega-3 fatty acids [1,2]. Despite their potential, concerns persist regarding potential toxicity at higher concentrations [3].

Various nanoemulsions, such as those loaded with soybean isoflavone aglycones-rich fraction and essential clove oil, have demonstrated wound-healing properties [4]. However, nanoemulsions containing fish oil, despite their reported anti-inflammatory and antibacterial effects, remain relatively underexplored [5]. Recent studies highlight the potential of nanoemulsions as carriers for bioactive compounds, enhancing their stability and bioavailability.

Our previous research has demonstrated the antibacterial and anti-inflammatory potential of NE-FLO™ [5]. Zebrafish, which share a 70% genomic similarity with humans [6], serve as an ideal model for toxicity testing, as depicted in Figure 1. Studies indicate that exposure to high dosages of fish oil nanoemulsions can induce oxidative stress and histopathological changes in zebrafish [7,8]. Notably, one study found that nanoemulsions containing lemon oil had minimal adverse effects on Mozambique tilapia development and behavior, suggesting their suitability for eco-friendly toxicity assessments [9]. Additionally, Swathy et al. [10] examined the effects of black pepper oil nanoemulsion on shrimp. Their study reported no significant mortality or developmental abnormalities, highlighting the potential of these nanoemulsions as a non-toxic testing medium. Brine shrimp provides a simpler and cost-effective model for preliminary toxicity screening, which is valuable in the early-stage testing of new substances.

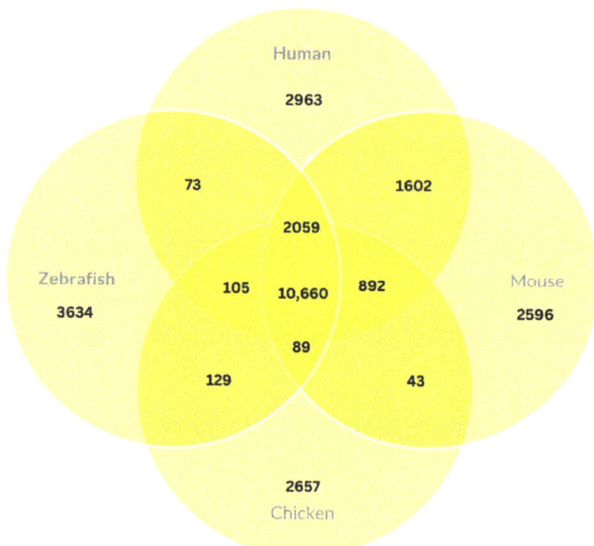

Figure 1. Orthologue genes shared between the zebrafish, human, mouse and chicken genomes, using orthology relationships from Ensembl Compara 63. Genes shared across species are considered in terms of copies at the time of the split. For example, a gene that exists in one copy in zebrafish but has been duplicated in the human lineage will be counted as only one shared gene in the overlap s reproduced from Howe et al. [6] Published by Nature, 2013, under Creative Commons Attribution-Non-Commercial-Share Alike license (http://creativecommons.org/licenses/by-nc-sa/3.0/).

The adult zebrafish toxicity test, known for its cost-effectiveness, reproducibility, and high throughput capacity, serves as a reliable alternative to traditional animal testing methods [11]. Zebrafish have been extensively used in toxicological research, providing valuable insights into chemical toxicity and aiding drug development [12,13].

Incorporating fish byproducts in the formulation of nanoemulsions provides a sustainable approach to waste reduction and adds value to fisheries. The effectiveness of fish oil-derived emulsions in delivering hydrophobic compounds without significant cellular toxicity has been reported [6]. This approach not only addresses waste management but also promotes the use of renewable resources to develop eco-friendly products. Therefore, this study aims to assess the toxicity profile of the NE-FLO™ optimized formulation to balance beneficial properties with minimal toxicity. This research contributes to the field of nanoemulsion science and holds practical implications for the pharmaceutical, nutraceutical, and food industries. Enhanced bioavailability of omega-3 fatty acids in these sectors can lead to better health outcomes and address dietary deficiencies on a larger scale.

2. Materials and Methods

2.1. Materials

This study used specific materials and equipment for experimental procedures. Brine shrimp cultivation involved *Artemia salina* eggs from Ocean Nutrition and pure *Artemia* cysts sourced from the Great Salt Lake, supplemented with sodium chloride (Sigma–Aldrich, catalog number: S7653, St. Louis, MO, USA) and sea salt (Tropic Marin, Hünenberg, Switzerland) in seawater conditions. Zebrafish (*Danio rerio*) maintenance utilized adult fish aged between 3 and 6 months from the Universiti Putra Malaysia (UPM) *Danio rerio* laboratory, housed in four 3 L fish tanks. Water quality was maintained using a Panasonic TK-CS20 water filter (Kadoma, Japan), supplemented with commercial fish feed Otohime B1 (Marubeni Nisshin Feed Co., Ltd., Tokyo, Japan), water conditioner using Nutrafin aqua plus (RC Hagen Ltd., Montreal, QC, Canada) and Nutrafin Clear (RC Hagen Ltd., Montreal, QC, Canada). Each tank was equipped with aquaria biochemical sponge foam filters to maintain water quality, and water parameters were monitored using an API freshwater master test kit. Other materials and consumables were used as per standard laboratory practices.

2.2. Ethical Clearance

Our research protocol was reviewed by the Institutional Animal Care and Use Committee (IACUC) and approved for animal use under the ID IACUC 2023-001 on 15 April 2023, valid until 15 April 2024. Animal experiments were completed in August 2023. Proper disposal procedures for zebrafish carcasses, as outlined in the IACUC protocol, were followed [14,15].

2.3. Preparation of the Nanoemulsion of Fish Byproducts and Lemon Oils (NE-FLO)™

The ultrasonication method utilized in this study was adapted from Azmi et al. [5] and Espinosa-Andrews and Paez-Hernandez [16]. The preparation of fish byproducts and lemon oil nanoemulsion (NE-FLO)™ involved a two-step process. First, the aqueous phase was created by stirring Tween 80 and the cosurfactant in pure distilled water for 15 min to form the continuous phase. Simultaneously, the oil phase, consisting of fish oil and lemon oil, was stirred at 500 rpm until thoroughly mixed (specific data are withheld due to patent processing) using a magnetic stirrer. Next, the oil phase was added dropwise into the continuous aqueous phase while maintaining the stirring speed. The mixture was homogenized using a T25 digital Ultra-Turrax homogenizer (Staufen, Germany) at 20,000 rpm for 3 min, resulting in a coarse oil-in-water emulsion. To achieve nanosized droplets, the coarse emulsion underwent ultrasonication for 42 min, an optimized duration for this study at 80% of the maximum output of 100 W. Following ultrasonication, the solution was tested to verify the formation of nanosized droplets. Figure 2 shows the process of preparing the nanoemulsion [5].

2.4. Properties of Nanoemulsion of Fish Byproducts and Lemon Oils (NE-FLO™)

The particle size, polydispersity index (PDI), and zeta potential of NE-FLO™ were measured using a particle size analyzer (Malvern Instruments, Malvern, UK) at 25 °C. The samples were diluted with deionized water at a 1:9 ratio to prevent backscattering. The refractive indices of the particles and water used were 1.54 and 1.33, respectively. Each measurement was performed three times [5].

2.5. Diluting NE-FLO™ for Toxicity on Brine Shrimp

NE-FLO™ was stored in the laboratory at 4 °C until the commencement of the experimental studies. Stock solutions of NE-FLO™ were prepared by diluting the NE-FLO™ samples with seawater to achieve six different concentrations: 10 $mg \cdot L^{-1}$, 1 $mg \cdot L^{-1}$, 0.1 $mg \cdot L^{-1}$, 0.01 $mg \cdot L^{-1}$, 0.005 $mg \cdot L^{-1}$, and 0.001 $mg \cdot L^{-1}$. The contents were vigorously vortexed for 20 s at 2000 rpm to ensure a uniform distribution. The initial dilution sample, referred to as 'stock 0', was prepared by mixing 0.10 g of NE-FLO™ in 10 mL of seawater with 1 mL of Tween 80% in 90 mL of seawater. Appropriate volumes of the stock solutions were then promptly transferred into the exposure containers containing *Artemia* in seawater. Figure 3 illustrates the steps required to prepare the brine shrimp for toxicity testing.

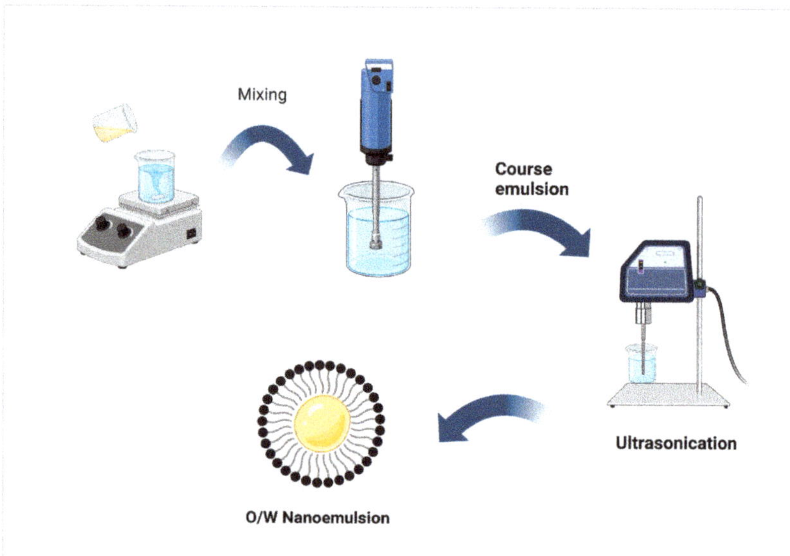

Figure 2. Schematic diagram of the preparation of NE-FLO™ using the ultrasonication method. The process involves the emulsification of lemon oil and fish byproduct oil, followed by ultrasonication to form a stable nanoemulsion. Reproduced with permission from Azmi et al. [5], Molecules; published by MDPI, 2022 under Creative Commons Attribution 4.0 International.

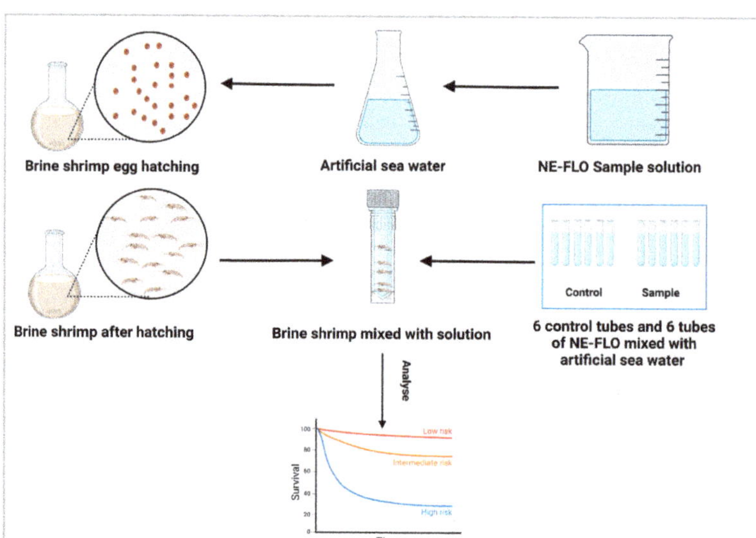

Figure 3. Preparation of the brine shrimp assay. Brine shrimp eggs hatch in artificial seawater over 24 h. They were then placed in various concentrations of NE-FLO™ and control samples to measure mortality rates.

2.6. NE-FLO™ Dilution for Adult Zebrafish

Stock solutions were prepared by diluting the nanoemulsion sample with filtered tap water to achieve concentrations of 1 mg·L^{-1}, 0.1 mg·L^{-1}, 0.01 mg·L^{-1}, and 0.001 mg·L^{-1}. These mixtures were vortexed for 20 s at 2000 rpm to ensure a uniform distribution. For instance, to create a 0.01 mg·L^{-1} concentration, a 5-L tank of clean fish tank water was prepared. The required volume of nanoemulsion treatment was calculated by using Equation (1).

$$\text{Volume} = \frac{\text{Desired concentration } \left(\text{mg}\cdot\text{mL}^{-1}\right) \times \text{Total tank volume (liters)}}{\text{Concentration of nanoemulsion treatment } \left(\text{mg}\cdot\text{mL}^{-1}\right)} \quad (1)$$

The calculated volume was then added to the tank and gently mixed to ensure thorough dispersion. Following this dilution process, the fish in the tank were observed for behavioral and health changes. The experimental protocol adhered to the Organization for Economic Cooperation and Development (OECD) guidelines for chemical testing [17].

2.7. Preparation of *Artemia salina* Larvae (Brine Shrimp)

The brine shrimp eggs (*Artemia salina*) and pure *Artemia* cysts were sourced from the Great Salt Lake. Geographical variations do not affect the assay. The cysts were prepared in-house and stored in a refrigerator at 4 °C. The incubated larvae hatched in artificial seawater (3% m/v) made by dissolving a suitable amount of Instant Ocean® salt (Blacksburg, VA, USA) in deionized water, stirring it for 24 h with aeration, and filtering it through Millipore cellulose filters. *Artemia* was hatched using a method described by Zhu B et al. [18], with some modifications. In brief, encysted *Artemia* was hydrated in distilled water for 12 h at 4 °C before being rinsed to remove floating cysts. Sinking cysts were collected using a Buchner funnel and rinsed in cold deionized water. Precleaned cysts (3 g) were incubated at 30 °C in 1.5 L saltwater in a conical plastic container with graduations. A fluorescent bulb provided continuous daylight at 60 W. An aquarium air pump supplied aeration through a hose flowing to the bottom of the hatching device. Under these conditions, *Artemia* larvae hatched within 24 h.

2.8. Preparation of Adult Zebrafish

Zebrafish were supplied by the UPM zebrafish laboratory and delivered to the desired location (INHART Laboratory @ KICT Building, International Islamic University Malaysia, Gombak campus). They were acclimatized to their new environment following the acclimatization protocol for three weeks [15,19]. The number of fish lost during acclimatization was about 10% of the total number of fish. The adult zebrafish, both male and female, aged between 3 and 6 months and weighing 0.45 ± 0.05 g, were exposed to the solution with and without the fish oil nanoemulsions in covered 5 L plastic tanks [20,21]. Each experimental and control group contained five fish, with a ratio of 1 g of fish weight to 1.8–2.0 L of water. All tank water parameters, including pH, high range pH, ammonia, nitrite, and nitrate, were kept stable to ensure scientific accuracy. Control groups were maintained in water at 25 ± 1 °C without the nanoemulsion treatment. The total hardness of the water was 13 N° dgH, and the light-dark regimen was 12 h of light and 12 h of darkness. The oxygen concentration was maintained above 60%, and the pH was kept between 6.5 and 7.0. The test was initiated by diluting the four concentrations of nanoemulsion (1 mg·L^{-1}, 0.1 mg·L^{-1}, 0.01 mg·L^{-1}, and 0.001 mg·L^{-1}) with fish tank water. A negative control tank, which received the same care except for the nanoemulsion treatment, was used to observe any changes or effects. Once the pH and other parameters (chlorine, nitrite, nitrate, ammonia) were tested and adjusted, five adult zebrafish were added to each tank. The fish were monitored every 24 h for three days for any signs of discomfort, illness, rejection, or mortality. The fish were fed 4% of body weight in food per day, as specified in the zebrafish care and maintenance protocol [22,23].

2.9. Experiment Setup for Artemia Larvae

Acute toxicity testing was conducted according to the guidelines established by OECD [17]. *Artemia* larvae were exposed to six different concentrations of NE-FLO™ for 24 h: 10 mg·L^{-1}, 1.0 mg·L^{-1}, 0.1 mg·L^{-1}, 0.01 mg·L^{-1}, 0.005 mg·L^{-1}, and 0.001 mg·L^{-1}, following OECD protocols. A positive control group (10 mg of tetracycline in 1 mL of seawater) without the test substance was also included. Exposures were conducted at 24 ± 2 °C in triplicate using 1.0 L conical plastic containers filled with 500 mL of saltwater, maintaining a total salinity of 2.9–3% m/v and a light regime of 16:8 h light: dark. No food was provided during exposure. Detailed experimental conditions, percentages, and log concentrations are presented in Table 1 and Figure 3. Following exposure, a portion of each group's *Artemia* was rinsed with seawater and immediately transferred to a clean seawater bath. All tests were conducted in duplicate to ensure the accuracy of results.

Table 1. Main properties of NE-FLO™.

Measurement	NE-FLO Nanoemulsion Value
Mean Particle Size (Z-averages)	44.40 ± 0.11 nm
Polydispersity Index (PDI)	0.077
Zeta Potential	−5.02 ± 0.22 mV
pH	4.27 ± 0.01

2.10. Experiment Design for Adult Zebrafish

A controlled laboratory environment was established for the adult zebrafish toxicity test with four separate fish tanks, each accommodating a group of five zebrafish. Each tank was subjected to a distinct concentration of the nanoemulsion treatment, while one tank served as the control with no additional substances introduced. These specific concentrations were determined based on the results of the brine shrimp toxicity test, taking into account the observed number of fatalities. Throughout the testing period, the zebrafish were fed a diet of commercial fish food, specifically Otohime B1 (Marubeni Nisshin Feed Co., Ltd., Tokyo, Japan), to meet their nutritional requirements. Environmental conditions were meticulously maintained, with temperature levels ranging between 26 and 28 °C and pH levels maintained within the range of 6.8 to 7.5. Figure 4 illustrates the laboratory setup employed during the adult zebrafish toxicity test.

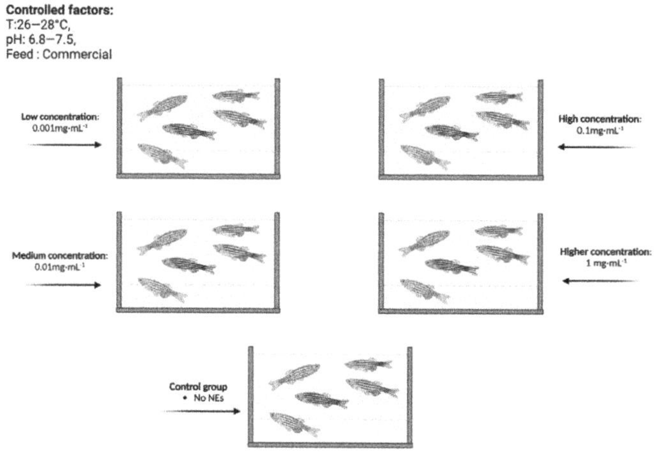

Figure 4. Illustration of the laboratory setup for the adult zebrafish toxicity test, displaying the arrangement of fish tanks. The tested concentrations, determined based on the brine shrimp test results, include low (0.001 mg·L^{-1}), medium (0.01 mg·L^{-1}), high (0.1 mg·L^{-1}), and a higher concentration (1 mg·L^{-1}) alongside a control tank. [NOTE: 1 ppm = 1 mg·L^{-1}].

2.11. Calculation of LC_{50}

The LC_{50} value for NE-FLO™ was determined using logistic regression analysis. Various concentrations of NE-FLO™ in mg/L were tested, and the mortality rates were recorded. The concentrations were log-transformed, and the mortality data were converted to proportions. A logistic regression model was then fitted to the data, with the log-transformed concentrations as the predictor variable. The LC_{50} value, representing the concentration at which 50% mortality is expected, was calculated from the model's parameters. The analysis was performed using the 'statsmodels' library in Python 3.11.10 (2024).

2.12. Statistical Analysis

In this study, statistical analyses were carried out using Microsoft Excel (Microsoft 365, version 2308).

2.13. OpenAI Utilization

In this study, a Python-based method was employed to calculate the LC_{50} value for NE-FLO™. Logistic regression analysis was performed on log-transformed concentrations and corresponding mortality data, with model fitting carried out using the stats models library in Python 3.11.10 (2024). ChatGPT was used to assist in developing and optimizing the code for the LC50 determination. Additionally, ChatGPT was utilized to edit and check the manuscript's English language for coherence and clarity. All outputs generated by ChatGPT were thoroughly reviewed and verified by the authors to ensure accuracy.

3. Results and Discussion

3.1. Properties of the Nanoemulsion of Fish Byproducts and Lemon Oils (NE-FLO™)

Table 1 shows the primary quantitative results of the NE-FLO nanoemulsion investigation. The average particle size (Z-averages) is 44.40 ± 0.11 nm, indicating a highly fine distribution of particles within the nanoemulsion. A low PDI value of 0.077 reflects a high consistency and uniformity of the nanoemulsion droplet size. The zeta potential recorded at -5.02 ± 0.22 mV indicates adequate electrostatic stabilization despite being below the ideal stability range (greater than $|30|$ mV), thereby mitigating any concerns of flocculation or sedimentation. At higher concentrations, instability may lead to aggregation, potentially increasing toxicity. Future studies should assess the stability of the nanoemulsion in brine water and evaluate the impact of any aggregation on toxicity. Lastly, the nanoemulsion's pH is 4.27 ± 0.01, being acidic, which may influence the zeta potential and overall formulation stability [5].

3.2. Artemia Larvae Toxicity Test Observations

The toxicity experiments were conducted in triplicates on brine shrimp. Each trial involved six different concentrations of brine shrimp media and fish oil nanoemulsions, as indicated in Table 2. The mortality rates are calculated according to the equation:

$$\text{Mortality }\% = \frac{\text{Shrimp No.(initial)}}{\text{Total Mortality}} \times 100. \quad (2)$$

In the saltwater media with concentrations of 0.001 $mg{\cdot}L^{-1}$, 0.005 $mg{\cdot}L^{-1}$, and 0.01 $mg{\cdot}L^{-1}$, the mortalities observed were five brine shrimps or lower. However, at higher concentrations (0.1 $mg{\cdot}L^{-1}$, 1 $mg{\cdot}L^{-1}$, and 10 $mg{\cdot}L^{-1}$), a consistent mortality rate of 100.0% was observed, indicating that all the brine shrimps in these concentrations died. This suggests that the brine shrimp could not survive at higher concentrations, indicating these levels are highly toxic. The increase in mortality at higher concentrations highlights the critical impact of dosage on brine shrimp viability.

Table 2. Results of the acute toxicity test on *Artemia* Larvae.

	Sample Conc. ($mg{\cdot}L^{-1}$)	Log Conc. ($mg{\cdot}L^{-1}$)	Replicate One				Replicate Two				Replicate Three			
			Shrimps No. (Initial)	Shrimps No. (Final)	Total Mortality	Mortality %	Shrimps No. (Initial)	Shrimps No. (Final)	Total Mortality	Mortality %	Shrimps No. (Initial)	Shrimps No. (Final)	Total Mortality	Mortality %
Positive Control *	0.001	−2.3	10	7	3	30.0%	10	7	3	30.0%	10	7	3	30.0%
	0.005	−1.8	10	6	4	40.0%	10	6	4	40.0%	10	6	6	60.0%
	0.01	−1.5	10	5	5	50.0%	10	5	5	50.0%	10	5	5	50.0%
	0.1	−0.8	10	0	10	100.0%	10	0	10	100.0%	10	0	10	100.0%
	1	0	10	0	10	100.0%	10	0	10	100.0%	10	0	10	100.0%
	10	0.8	10	0	10	100.0%	10	0	10	100.0%	10	0	10	100.0%
Sample **	0.001	−2.3	10	8	2	20.0%	10	8	2	20.0%	10	8	2	20.0%
	0.005	−1.8	10	7	3	30.0%	10	7	3	30.0%	10	7	3	30.0%
	0.01	−1.5	10	7	3	30.0%	10	6	4	40.0%	10	7	3	30.0%
	0.1	−0.8	10	6	4	40.0%	10	6	4	40.0%	10	6	4	40.0%
	1	0	10	5	5	50.0%	10	4	6	60.0%	10	6	4	40.0%
	10	0.8	10	4	6	60.0%	10	3	7	70.0%	10	4	6	60.0%
Negative Control * (Artificial Seawater)**	**Replicate 1**	10	9	1		10								
	Replicate 2	10	8	2		20								
	Replicate 3	10	9	1		10								
	Average total death			1.33										

* Positive control: Diluted tetracycline in seawater added to the brine shrimp with concentrations of 0.001 $mg{\cdot}L^{-1}$, 0.005 $mg{\cdot}L^{-1}$, 0.01 $mg{\cdot}L^{-1}$, 0.1 $mg{\cdot}L^{-1}$, 1 $mg{\cdot}L^{-1}$, and 10 $mg{\cdot}L^{-1}$. ** Sample: Diluted NE-FLO™ treatment added to the brine shrimp with concentrations of 0.001 $mg{\cdot}L^{-1}$, 0.005 $mg{\cdot}L^{-1}$, 0.01 $mg{\cdot}L^{-1}$, 0.1 $mg{\cdot}L^{-1}$, 1 $mg{\cdot}L^{-1}$, and 10 $mg{\cdot}L^{-1}$. *** Negative control: Artificial seawater as a medium for the brine shrimp.

Similarly, mortality rates were calculated for the NE-FLO™ treatments at the same concentrations. It was observed that the concentrations of 0.001 mg·L^{-1}, 0.005 mg·L^{-1}, and 0.01 mg·L^{-1} resulted in 2 to 3 brine shrimp mortalities, while the concentrations of 0.1 mg·L^{-1}, 1 mg·L^{-1}, and 10 mg·L^{-1} resulted in 4 to 6 mortalities. Specifically, the concentration of 0.1 mg·L^{-1} exhibited the lowest mortality rate at 40.0%. The concentration of 1 mg·L^{-1} had a mortality rate of 50.0%. Lastly, the concentration of 10 mg·L^{-1} exhibited a 60.0% mortality rate. These percentages reflect the proportion of mortalities relative to the total number of samples tested at each concentration.

At the 48 h mark, live brine shrimp counts were recorded again, and the mortality rates for each concentration remained consistent with the first trial except for a slight rise in the mortality rates for the concentrations of 1 mg·L^{-1} and 10 mg·L^{-1} where the mortality rates are observed to be 60% and 70%, respectively. After 72 h, it was observed that higher concentrations of diluted tetracycline in seawater (0.1 mg·L^{-1}, 1 mg·L^{-1}, and 10 mg·L^{-1}) resulted in a 100% mortality rate among the brine shrimps. In contrast, fish oil nanoemulsion treatments at equivalent concentrations showed lower mortality rates, ranging between 40.0% and 60.0%. Specifically, concentrations resulted in 3 to 4 mortalities, signifying a comparatively higher survival rate compared to the saltwater media concentrations.

As observed from Table 2, the acute toxicity test results for brine shrimp across various concentrations provide an understanding of the toxic effects of the tested sample. At lower concentrations, the sample showed relatively low toxicity to brine shrimp. However, as concentrations increased, a marked rise in mortality rates was observed. The control group at 0.005 mg·L^{-1} and 0.01 mg·L^{-1} exhibited mortality rates of 40% and 50%, respectively, 26.67%, and maintained consistency, indicating a threshold where adverse effects become noticeable. Interestingly, the sample group at 0.01 mg·L^{-1} displayed similar mortality rates, suggesting a comparable level of toxicity to the control at this concentration.

A significant increase in mortality is observed at 0.1 mg·L^{-1}, with both control and sample groups experiencing a 100% mortality rate, indicating a concentration-dependent response where brine shrimp are unable to survive higher doses. This finding highlights the concentration level at which the sample exhibits acute toxicity, consistent with the effects seen in the control substance. At the highest concentrations (1 mg·L^{-1} and 10 mg·L^{-1}), positive control and sample groups maintained high mortality rates at 100%. However, the sample group showed slight variation with mortality rates of 50% and 60%, respectively, indicating a possible variation in the lethal dose between replicates.

Such variability emphasizes the influence of environmental or procedural factors such as temperature, water quality, and contaminants, which can significantly impact the toxicity of substances to aquatic organisms. For instance, variations in water temperature and wastewater effluent compositions have been shown to affect the results of toxicity bioassays [24]. Hence, careful monitoring and maintenance of water quality during testing are crucial to minimize these variables. The negative control group treated with artificial seawater displayed significantly lower mortality rates, with an average total death of only 1.33. This stark contrast indicates that the observed mortalities in the other groups are directly attributable to the concentrations of the tested substances rather than environmental conditions.

Our results suggest that the fish oil nanoemulsion treatment exhibited relatively low mortality rates even at higher concentrations compared to the artificial seawater brine shrimp media and the positive control group. The selected concentrations for the zebrafish tests (0.1 mg·L^{-1}, 0.01 mg·L^{-1}, and 0.001 mg·L^{-1}) were chosen to account for potential differences in sensitivity between brine shrimp and larger organisms like zebrafish. Brine shrimp may respond differently to the tested concentrations compared to zebrafish [25]. To further assess these differences, higher concentrations (1 mg·L^{-1} and 10 mg·L^{-1}) were also tested on zebrafish, allowing for the observation of effects without reaching the high mortality rates seen at the maximum concentrations in the brine shrimp tests, which led to 100% mortality rates. Figure 5 illustrates the relationship between the NE-FLO™

sample's log concentration and the mortality rate. The drastic change in zebrafish livability observed between 0.1 mg·L^{-1} and 1 mg·L^{-1} of NE-FLO could be attributed to a limit effect, where a critical concentration is reached, causing significant physiological stress. At lower concentrations (e.g., 0.1 mg·L^{-1}), the nanoemulsion may not have reached a level sufficient to disrupt cellular processes or provoke a toxic response. However, at higher concentrations (e.g., 1 mg·L^{-1}), the accumulation of the nanoemulsion may surpass the organism's capacity to cope with stress, leading to mortality.

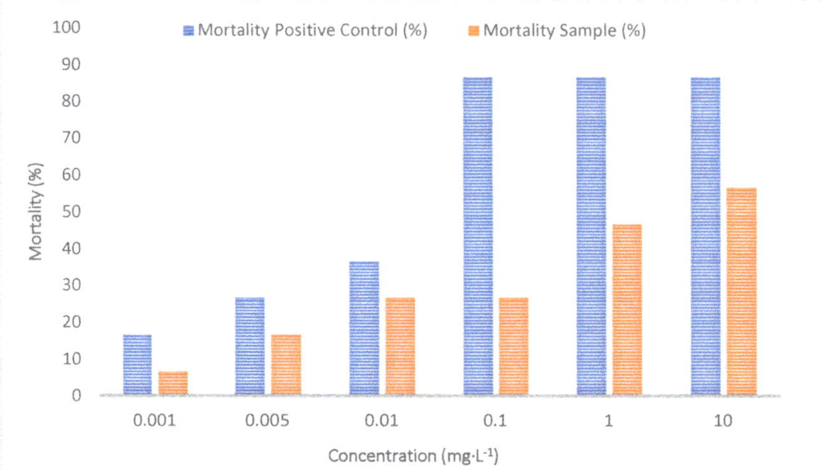

Figure 5. Effect of NE−FLO™ concentration on brine shrimp mortality percentage.

The logistic regression analysis estimated the LC_{50} value for the tested substance to be 8.7474 mg·L^{-1}, indicating moderate toxicity to brine shrimp. This value suggests that the substance poses a moderate risk to aquatic environments where brine shrimp are present. While the current model provides a useful initial estimate of the LC_{50}, further studies with larger sample sizes and a wider range of concentrations would enhance the robustness and precision of these findings. Consequently, it is advisable to exercise caution when using this substance in relevant environments to mitigate potential impacts on brine shrimp.

3.3. Adult Zebrafish Toxicity Test Observation

On day zero of the zebrafish acute toxicity test, no mortality was observed, and all the fish exhibited signs of good health. This was evident from their normal swimming patterns and responsive behavior towards external stimuli in all three tanks treated with fish oil nanoemulsion. The fish were fed according to recommended guidelines, and no decrease in appetite was observed. Additionally, their excretions appeared normal in color and quantity compared to the zebrafish observed in the negative control tank (Figure 6). In contrast, zebrafish exposed to higher concentrations of toxic substances often display abnormal swimming patterns, such as erratic or hyperactive swimming, circling or looping behavior, or even immobility. Loss of equilibrium, where fish swim on their sides or upside down, is another sign of severe distress. These abnormal behaviors have been well-documented as indicators of neurotoxicity or stress in zebrafish models of environmental toxicity. None of these abnormal behaviors were observed in the fish oil nanoemulsion treatment groups, further indicating the safety and nontoxicity of the sample at the tested concentrations [26].

After a 24 h period, the zebrafish were reevaluated. No mortality was recorded at concentrations of 0.1 mg·L^{-1}, 0.01 mg·L^{-1}, and 0.001 mg·L^{-1}, indicating 0% mortality for these concentrations. Zebrafish in these tanks exhibited normal swimming behaviors comparable to those in the negative control tank, and the feeding regimen continued without any issues. However, at the concentration of 1.0 mg·L^{-1}, a 100% mortality rate was observed, with all five fish found dead. This result indicates that adult zebrafish could not tolerate the 1.0 mg·L^{-1} concentration.

Figure 6. Zebrafish exposed to a 0.1 mg·L^{-1} concentration of NE-FLO™, displaying normal swimming patterns and responsive behavior towards external stimuli.

On the second day (48 h), observations remained consistent with those of day one. Concentrations of 0.1 mg·L^{-1}, 0.01 mg·L^{-1}, and 0.001 mg·L^{-1} continued to show 0% mortality, with zebrafish exhibiting normal living and swimming behaviors. By the third day (72 h), the final observations were similar to those of the previous days. Concentrations of 0.1 mg·L^{-1}, 0.01 mg·L^{-1}, and 0.001 mg·L^{-1} showed 0% mortality, and the fish remained alive and in good health, with no negative changes in swimming patterns, appetite, or excretions. Throughout the three days, the negative control tank showed no fish deaths, confirming that the water conditions did not contribute to mortality.

The estimated LC_{50} value for the tested emulsion is 0.316 mg·L^{-1}, indicating a high level of toxicity to adult zebrafish. According to standard toxicity classifications, substances with an LC_{50} value of less than 1 mg·L^{-1} are considered highly toxic. Despite this classification, the substance exhibited no toxicity at lower concentrations, with mortality occurring only at higher concentrations. This pattern suggests that the substance can be used safely at lower concentrations without posing significant risks to aquatic environments. However, at higher concentrations, the substance can cause significant mortality, necessitating stringent measures to manage and mitigate potential environmental impacts, especially in habitats supporting sensitive aquatic species. Further studies are recommended to explore the mechanisms of toxicity at higher concentrations and to assess the long-term ecological effects of the substance. Table 3 and Figure 7 illustrate a detailed overview of the acute toxicity test duration, results, log concentration (mg·mL^{-1}), and viability percentage.

Gu et al. [27] and Achenbach et al. [28] suggest investigating how toxicity operates at higher concentrations. Understanding whether the sample induces mortality by damaging specific organs, disrupting metabolic processes, or triggering other physiological stress responses can aid in developing better strategies to mitigate risks and establish safer application methods. Detailed studies, such as those on the heart-related effects of doxorubicin in zebrafish, illustrate how specific effects can be uncovered and potentially addressed [29].

Table 3. Results of the acute toxicity test on adult zebrafish.

	Sample Conc. ($mg.mL^{-1}$)	Log Conc. ($mg.mL^{1}$)	Day One (24 h)				Day Two (48 h)				Day Three (72 h)			
			No. of Fish (Initial)	No. of Fish (Final)	Total Deaths	Mortality %	No. of Fish (Initial)	No. of Fish (Final)	Total Deaths	Mortality %	No. of Fish (Initial)	No. of Fish (Final)	Total Deaths	Mortality %
Sample *	0.001	−2.3	5	5	0	0	5	5	0	0	5	5	0	0
	0.01	−1.5	5	5	0	0	5	5	0	0	5	5	0	0
	0.1	−0.8	5	5	0	0	5	5	0	0	5	5	0	0
	1	0.0	5	0	5	100								
Negative Control **	Filtered tap water		5	5	0									
	Average total deaths				0									

* Sample: Diluted NE-FLO™ treatment added to the brine shrimp with concentrations 0.001 $mg{\cdot}L^{-1}$, 0.01 $mg{\cdot}L^{-1}$, 0.1 $mg{\cdot}L^{-1}$, and 1 $mg{\cdot}L^{-1}$. ** Negative control: Filtered and conditioned tap water was used as a medium for the zebrafish.

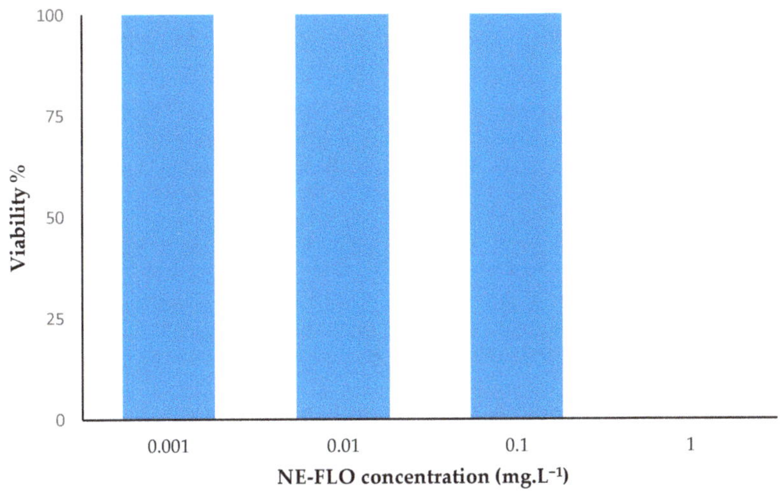

Figure 7. Relation between adult zebrafish viability percentage and NE−FLO™ concentration.

Zain et al. [30] reported on the use of flavonoid-enriched oil palm (*Elaeis guineensis* Jacq.) leaf extract (OPL) using the zebrafish model. Their study aimed to develop a flavonoid-enriched extract of OPL (OPL-FEE) and incorporate it into a nanoemulsion system (OPL-FEE-NE) to assess its toxicity, wound healing properties, and transcriptional effects using the zebrafish model. Based on embryonic toxicity assay results, they established test concentrations for the exposure experiments. Zebrafish mortality rates were monitored at 24, 48, 72, and 96 h, revealing a dose-dependent effect for the test samples, consistent with previous studies [31]. OPL-FEE exhibited high toxicity to adult zebrafish at 100 mg·L^{-1}, while OPL-FEE-NE showed toxicity at concentrations above 2 g·L^{-1}, resulting in 100% mortality within 24 h of exposure. However, after 96 h of exposure, no mortality was observed at test concentrations of 25 mg·L^{-1} for OPL-FEE and 0.5 g·L^{-1} for OPL-FEE-NE, indicating the safety of these concentrations for subsequent wound healing experiments.

4. Conclusions

The acute toxicity assessments conducted on brine shrimp and adult zebrafish followed appropriate protocols to ensure safety across a range of NE-FLO™ concentrations. These tests revealed a favorable safety profile, with minor mortalities observed in the brine shrimp toxicity test for 0.001 mg·L^{-1}, 0.005 mg·L^{-1}, and 0.01 mg·L^{-1} concentrations with increased mortality rates for higher concentrations of 0.1 mg·L^{-1}, 1.0 mg·L^{-1}, and 10 mg·L^{-1} in which the size of the organisms is taken into consideration. Similarly, no mortalities or adverse effects were observed in the acute toxicity evaluation on adult zebrafish at concentrations of 0·1 mg·L^{-1}, 0.01 mg·L^{-1}, and 0.001 mg·L^{-1}.

Additionally, the LC_{50} values for the nanoemulsion were 8.7474 mg·L^{-1} for brine shrimp and 0.316 mg·L^{-1} for adult zebrafish. They exhibited no toxicity at lower concentrations but caused mortality at higher ones, suggesting safe usage at lower doses and potential risks at higher doses, especially in sensitive aquatic environments.

These findings affirm the overall safety of fish oil nanoemulsions in acute toxicity tests, indicating potential for diverse industrial applications. This study lays a solid groundwork for further exploring nanoemulsions as safe and effective carriers for bioactive compounds. Nevertheless, continued studies are necessary to fully evaluate their safety and efficacy across broader applications and prolonged exposure durations.

5. Patents

This work was generated from the patent application of the formulation NE-FLO™, which is currently in the First Substantive Examination Adverse Report, Malaysian Patent Application PI 2022004802.

Author Contributions: A.A.H. performed the experiment, analyzed the data, validated the result, and wrote the manuscript. N.A.N.A. performed the nanoemulsion experiments. A.A.M.E. and N.A.-S. designed the research, obtained the funds, and co-supervised the first author. N.A.-S., H.M.S. and A.A.M.E., provided research resources, analyzed the data, and reviewed the manuscript. N.M.N. contributed to analyzing the data and editing the manuscript. All authors have read and agreed to the published version of the manuscript.

Funding: This work is financially aided by the Ministry of Higher Education, Malaysia, under the Fundamental Research Grant Scheme granted to Asst. Prof. Dr. Nurhidayu Al-saari [FRGS/1/2019/WAB01/UIAM/02/1] and [SPD23-002-0002] to Asst. Prof. Dr. Amal A. M. Elgharbawy.

Institutional Review Board Statement: This study was conducted in accordance with the Declaration of IIUM Animal Care and Use Committee (I-ACUC), Research Management Center (RMC), Kuantan Campus, and approved by the Ethical Committee.

Data Availability Statement: Some data presented in this study are available upon request from the corresponding author due to patent processing.

Acknowledgments: We extend our sincere gratitude to the International Islamic University Malaysia (IIUM) and the International Institute for Halal Research and Training (INHART) for their invaluable support. The authors acknowledge the use of OpenAI during the preparation of this work to enhance the writing. After utilizing the tool, the authors thoroughly reviewed and edited the content as necessary and assumed full responsibility for the publication's content.

Conflicts of Interest: The authors declare no conflicts of interest.

References

1. Back, P.I.; Balestrin, L.A.; Fachel, F.N.S.; Nemitz, M.C.; Falkembach, M.; Soares, G.; Marques, M.d.S.; Silveira, T.; Prá, M.D.; Horn, A.P.; et al. Hydrogels containing soybean isoflavone aglycones-rich fraction-loaded nanoemulsions for wound healing treatment—In vitro and in vivo studies. *Colloids Surf. B Biointerfaces* **2020**, *196*, 111301. [CrossRef] [PubMed]
2. Yadav, N.; Parveen, S.; Banerjee, M. Potential of nano-phytochemicals in cervical cancer therapy. *Clin. Chim. Acta* **2020**, *505*, 60–72. [CrossRef] [PubMed]
3. Santos, D.S.; Morais, J.A.V.; Vanderlei, A.; Santos, A.S.; Azevedo, R.B.; Muehlmann, L.A.; Júnior, O.R.; Mortari, M.R.; da Silva, J.R.; da Silva, S.W.; et al. Oral delivery of fish oil in oil-in-water nanoemulsion: Development, colloidal stability and modulatory effect on in vivo inflammatory induction in mice. *Biomed. Pharmacother.* **2021**, *133*, 110980. [CrossRef] [PubMed]
4. Banerjee, K.; Madhyastha, H.; Sandur, R.; Manikandanath, N.T.; Thiagarajan, N.; Thiagarajan, P. Anti-inflammatory and wound healing potential of a clove oil emulsion. *Colloids Surf. B Biointerfaces* **2020**, *193*, 111102. [CrossRef] [PubMed]
5. Azmi, N.A.N.; Elgharbawy, A.A.M.; Salleh, H.M.; Moniruzzaman, M. Preparation, Characterization and Biological Activities of an Oil-in-Water Nanoemulsion from Fish Byproducts and Lemon Oil by Ultrasonication Method. *Molecules* **2022**, *27*, 6725. [CrossRef]
6. Howe, K.; Clark, M.D.; Torroja, C.F.; Torrance, J.; Berthelot, C.; Muffato, M.; Collins, J.E.; Humphray, S.; McLaren, K.; Matthews, L.; et al. The zebrafish reference genome sequence and its relationship to the human genome. *Nature* **2013**, *496*, 498–503. [CrossRef]
7. Shehzad, Q.; Rehman, A.; Jafari, S.M.; Zuo, M.; Khan, M.A.; Ali, A.; Khan, S.; Karim, A.; Usman, M.; Hussain, A.; et al. Improving the oxidative stability of fish oil nanoemulsions by co-encapsulation with curcumin and resveratrol. *Colloids Surf. B Biointerfaces.* **2021**, *199*, 111481. [CrossRef]
8. Inapurapu, S.P.; Ibrahim, A.; Kona, S.R.; Pawar, S.C.; Bodiga, S.; Bodiga, V.L. Development and characterization of ω-3 fatty acid nanoemulsions with improved physicochemical stability and bioaccessibility. *Colloids Surf. A Physicochem. Eng. Asp.* **2020**, *606*, 125515. [CrossRef]
9. Baba, E.; Acar, Ü.; Öntaş, C.; Kesbiç, O.S.; Yılmaz, S. Evaluation of Citrus limon peels essential oil on growth performance, immune response of Mozambique tilapia Oreochromis mossambicus challenged with Edwardsiella tarda. *Aquaculture* **2016**, *465*, 13–18. [CrossRef]
10. Swathy, J.S.; Mishra, P.; Thomas, J.; Mukherjee, A.; Chandrasekaran, N. Antimicrobial potency of high-energy emulsified black pepper oil nanoemulsion against aquaculture pathogen. *Aquaculture* **2018**, *491*, 210–220. [CrossRef]
11. Chahardehi, A.M.; Arsad, H.; Lim, V. Zebrafish as a Successful Animal Model for Screening Toxicity of Medicinal Plants. *Plants* **2020**, *9*, 1345. [CrossRef] [PubMed]

12. Kovrižnych, J.A.; Sotníková, R.; Zeljenková, D.; Rollerová, E.; Szabová, E.; Wimmerová, S. Acute toxicity of 31 different nanoparticles to zebrafish (*Danio rerio*) tested in adulthood and in early life stages—Comparative study. *Interdiscip. Toxicol.* **2013**, *6*, 67–73. [CrossRef] [PubMed]
13. Caballero, M.V.; Candiracci, M. Zebrafish as Toxicological model for screening and recapitulate human diseases. *J. Unexplored Med. Data.* **2018**, *3*, 4. [CrossRef]
14. Institutional Animal Care and Use Committee. *IACUC Policy 23: Use of Zebrafish for Research and Teaching*; Institutional Animal Care and Use Committee: Chicago, IL, USA, 2018; p. 2.
15. Gilbert, M.J.; Adams, O.A.; Farrell, A.P. A sudden change of heart: Warm acclimation can produce a rapid adjustment of maximum heart rate and cardiac thermal sensitivity in rainbow trout. *Curr. Res. Physiol.* **2022**, *5*, 179–183. [CrossRef] [PubMed]
16. Espinosa-Andrews, H.; Páez-Hernández, G. Optimization of ultrasonication curcumin-hydroxylated lecithin nanoemulsions using response surface methodology. *J. Food Sci. Technol.* **2020**, *57*, 549–556. [CrossRef]
17. OECD. *Test No. 203: Fish, Acute Toxicity Test. No. 203*; OECD Publishing: Paris, France, 2019; p. 24. [CrossRef]
18. Zhu, B.; Zhu, S.; Li, J.; Hui, X.; Wang, G.-X. The developmental toxicity, bioaccumulation and distribution of oxidized single walled carbon nanotubes in: Artemia salina. *Toxicol. Res.* **2018**, *7*, 897–906. [CrossRef]
19. The Organization for Economic Co-Operation and Development [OECD]. *OECD Guideline for Testing of Chemicals (Draft Revised Version)*, 11th ed.; The Organization for Economic Co-operation and Development: Paris, France, 2018.
20. Institutional Animal Care and Use Program. Zebrafish Care Standards. UCSF. 2022. Available online: https://iacuc.ucsf.edu/sites/g/files/tkssra751/f/wysiwyg/POLICY%20-%20Zebrafish%20Care%20Standards.pdf (accessed on 21 October 2024).
21. OECD. Guidance Document on Aquatic Toxicity Testing of Difficult Substances and Mixtures. In *Organisation for Economic Co-Operation and Development*, 2nd ed.; Series on Testing and Assessment No. 23; OECD Publishing: Paris, France, 2019; Volume 23, pp. 1–81. Available online: https://www.oecd-ilibrary.org/environment/detailed-review-paper-drp-on-molluscs-life-cycle-toxicity-testing_9789264221468-en (accessed on 21 October 2024).
22. Zulfahmi, I.; El Rahimi, S.A.; Suherman, S.D.; Almunawarah, A.; Sardi, A.; Helmi, K.; Nafis, B.; Perdana, A.W.; Adani, K.H.; Nasution, I.A.A.; et al. Acute toxicity of palm oil mill effluent on zebrafish (Danio rerio Hamilton-Buchanan, 1822): Growth performance, behavioral responses and histopathological lesions. *Chemosphere* **2023**, *340*, 139788. [CrossRef]
23. Claro, R.O.; Rivero-Wendt, C.L.G.; Miranda-Vilela, A.L.; Grisolia, C.K.; Facco, G.G.; de Lima Moreira, D.; de Fátima Guilhermino, J.; Matias, R. Toxicological effects of aqueous extract of *Genipa americana* L. leaves on adult zebrafish (Danio rerio): Chemical profile, histopathological effects and lack of genotoxicity. *Toxicon* **2023**, *235*, 107305. [CrossRef]
24. Luan, X.; Liu, X.; Fang, C.; Chu, W.; Xu, Z. Ecotoxicological effect of disinfected wastewater effluent: A short review of in vivo toxicity bioassays to aquatic organisms. *Environ. Sci.* **2020**, *6*, 2275–2286. [CrossRef]
25. Ortiz-Román, M.I.; Casiano-Muñiz, I.M.; Román-Velázquez, F.R. Toxicity of UV Filter Benzophenone-3 in Brine Shrimp Nauplii (*Artemia salina*) and Zebrafish (*Danio rerio*) Embryos. *J. Xenobiot.* **2024**, *14*, 537–553. [CrossRef]
26. Ford, A.T.; Ågerstrand, M.; Brooks, B.W.; Allen, J.; Bertram, M.G.; Brodin, T.; Dang, Z.; Duquesne, S.; Sahm, R.; Hoffmann, F.; et al. The Role of Behavioral Ecotoxicology in Environmental Protection. *Environ. Sci. Technol.* **2021**, *55*, 5620–5628. [CrossRef]
27. Gu, J.; Guo, L.; Chen, C.; Ji, G.; Wang, L. Neurobehavioral toxic effects and mechanisms of 2-aminobenzothiazole exposure on zebrafish. *Sci. Total Environ.* **2024**, *913*, 169495. [CrossRef] [PubMed]
28. Achenbach, J.C.; Leggiadro, C.; Sperker, S.A.; Woodland, C.; Ellis, L.D. Comparison of the Zebrafish Embryo Toxicity Assay and the General and Behavioral Embryo Toxicity Assay as New Approach Methods for Chemical Screening. *Toxics* **2020**, *8*, 4. [CrossRef] [PubMed]
29. Lane, S.; More, L.A.; Asnani, A. Zebrafish models of cancer therapy-induced cardiovascular toxicity. *J. Cardiovasc. Dev. Dis.* **2021**, *8*, 1–12. [CrossRef]
30. Zain, M.S.C.; Edirisinghe, S.L.; Kim, C.-H.; De Zoysa, M.; Shaari, K. Nanoemulsion of flavonoid-enriched oil palm (*Elaeis guineensis* Jacq.) leaf extract enhances wound healing in zebrafish. *Phytomed. Plus* **2021**, *1*, 100124. [CrossRef]
31. Richardson, R.; Slanchev, K.; Kraus, C.; Knyphausen, P.; Eming, S.; Hammerschmidt, M. Adult zebrafish as a model system for cutaneous wound-healing research. *J. Investig. Dermatol.* **2013**, *133*, 1655–1665. [CrossRef]

colloids and interfaces

MDPI

Article

Preparation of Colloidal Silver Triangular Nanoplates and Their Application in SERS Detection of Trace Levels of Antibiotic Enrofloxacin

Cao Tuan Anh [1], Dao Tran Cao [2,3] and Luong Truc-Quynh Ngan [2,*]

1 Department of Science, Technology and Environment, Ministry of Education and Training, 35 Dai Co Viet, Hanoi 100000, Vietnam; ctanh@moet.gov.vn
2 Institute of Materials Science, Vietnam Academy of Science and Technology, 18 Hoang Quoc Viet, Hanoi 100000, Vietnam; dtcao@ims.vast.ac.vn
3 Department of Materials Science and Energy, Graduate University of Science and Technology, Vietnam Academy of Science and Technology, 18 Hoang Quoc Viet, Hanoi 100000, Vietnam
* Correspondence: nganltq@ims.vast.ac.vn

Abstract: Surface-enhanced Raman scattering (SERS) is a powerful technique for detecting trace amounts of chemicals due to its capacity to significantly amplify the Raman signal of the molecules of these substances. This is particularly relevant in food systems where monitoring antibiotic residues is critical for food safety. Traditional SERS substrates typically utilize colloidal silver nanospheres (AgNSs), but anisotropic silver nanoparticles with numerous sharp tips can further enhance SERS sensitivity, enabling lower detection limits suitable for food safety regulations. In this study, we describe a straightforward synthesis of colloidal silver triangular nanoplates (AgTNPls), featuring multiple sharp tips, using only four common reagents: silver nitrate, trisodium citrate, sodium borohydride ($NaBH_4$) and hydrogen peroxide (H_2O_2), all at room temperature. By carefully controlling the sequence of reagent addition, specifically introducing H_2O_2 after $NaBH_4$, we achieved a two-step synthesis. In the first step, AgNSs seeds form, and in the second, these seeds convert into AgTNPls, resulting in a colloid of relatively uniform AgTNPls with an edge length of approximately 52 nm. The resulting AgTNPls colloid, combined with an aluminum foil, produced an SERS substrate with high enhancement factor of 3.2×10^9 (using rhodamine 6G as a test molecule). Applied to enrofloxacin (an antibiotic widely used in livestock and aquaculture) detection, this substrate achieved a detection limit as low as 0.39 μg/L (0.39 ppb), with enrofloxacin detectable at concentrations down to 5 μg/L. This highly sensitive SERS substrate holds great promise for rapid, accurate detection of antibiotic residues in food products, aiding regulatory compliance and food safety assurance.

Keywords: SERS; silver triangular nanoplates; antibiotic; enrofloxacin

1. Introduction

Surface-enhanced Raman scattering (SERS) is a powerful technique capable of amplifying Raman scattering signals from molecules that positioned near nanoscale rough metal surfaces, often achieving enhancement factors in the millions [1]. This remarkable amplification allows for the detection of trace concentrations of organic and biological molecules, making SERS highly effective for chemical and biochemical analyses.

The enhancement in SERS arises from the excitation of localized surface plasmons on nano-rough metal surfaces upon illumination. These plasmons generate intense electric fields that amplify the Raman scattering signals of molecules adsorbed on the surface. The extent of enhancement is influenced by the properties of the nano-rough surface, such as shape, size, and surface morphology. As a result, the preparation of optimal and stable nano-rough surfaces is crucial for achieving high SERS sensitivity.

Early SERS experiments employed electrochemically roughened silver surfaces [2], but advancements in nanotechnology have shifted the focus to designing substrates with metal nanoparticles distributed on solid supports [3]. The shape, size, spacing and arrangement of these metal nanoparticles significantly influence the enhancement factor (EF) of the Raman signal, as "hot spots"—regions of intense signal amplification—are typically found at the sharp edges, tips or in the narrow gaps between particles. To maximize enhancement, researchers have increasingly moved from isotropic spherical nanoparticles to anisotropic structures, such as triangular nanoplates, nanostars and flower-like particles, which feature sharp tips and edges.

Triangular nanoplates, or nanoprisms, are particularly attractive for SERS applications due to their well-defined sharp tips and edges that contribute to enhanced hot spots formation. Among the metals used for SERS, silver is favored over gold for its superior plasmonic activity [4–7]. Consequently, the synthesis and application of silver triangular nanoplates (AgTNPls) as SERS substrates have been extensively studied [8–14]. Two common methods for synthesizing AgTNPls are the photo-mediated and seed-mediated approaches. The photo-mediated approach involves light irradiation to transform spherical silver nanoparticles into triangular nanoplates [15], but it is often time-consuming and requires high temperatures. In contrast, the seed-mediated approach is more efficient and versatile [16], making it the method of choice for this study.

The seed-mediated synthesis of AgTNPls, first introduced by Mirkin's group [17], has undergone various refinements by researchers to improve efficiency and control [9–11,18–22]. In the original process, sodium borohydride ($NaBH_4$) was rapidly injected into a solution containing silver nitrate ($AgNO_3$), trisodium citrate (TSC), polyvinylpyrrolidone (PVP) and hydrogen peroxide (H_2O_2) at room temperature. To simplify the process, we removed PVP from the reactants and reversed the order of $NaBH_4$ and H_2O_2, based on their distinct functions: $NaBH_4$ reduces $AgNO_3$ to silver nanoparticles, while H_2O_2 facilitates their transformation into AgTNPls [23–26]. This modified approach, along with careful optimization of reactant concentrations, produced a colloid predominantly containing uniform AgTNPls, which was not achievable using the conventional method.

While colloidal AgTNPls can serve as SERS substrates, their instability due to nanoparticle movement limits their practical application. Solid substrates, prepared by immobilizing AgTNPls onto a solid support such as a glass slide or silicon wafer [10–13,27,28], offer greater stability and reproducibility. Recent studies have demonstrated that aluminum foil, due to its natural oxide layer (Al_2O_3), can enhance the local electromagnetic field via dielectric–metal interactions. The oxide layer acts as a dielectric spacer, preventing charge transfer quenching and supporting the formation of strong electromagnetic "hot spots" at the interface between AgTNPls and the Al substrate. Additionally, the rough surface morphology of commercial Al foil can facilitate nanoparticle immobilization and further intensify local plasmon coupling [29–35]. In this study, we found that AgTNPls immobilized on Al-foil exhibited enhancement factors hundreds of times greater than those on glass slides or silicon wafers.

Enrofloxacin (ENR), a fluoroquinolone antibiotic, is widely used in livestock and aquaculture due to its broad spectrum of antibacterial activity. However, its overuse has contributed to the rise in antibiotic-resistant bacteria and has raised concerns about harmful residues in food products [36]. Many countries have established maximum residue limits (MRLs) for ENR, typically ranging from 100 to 300 μg/kg, depending on the animal species and target tissue [37]. Accurate and sensitive detection of antibiotic residues is critical for ensuring food safety.

Conventional analytical methods for trace detection of antibiotics, including ENR, such as high-performance liquid chromatography (HPLC) [38,39] and liquid chromatography-mass spectrometry (LC/MS) [40,41], provide high sensitivity, accuracy and repeatability, but are time-consuming, require complex sample preparation and rely on expensive instrumentation. SERS offers a promising alternative, combining high sensitivity with simplicity, rapid analysis and minimal sample preparation. Additionally, as a molecular "fingerprint" technique, SERS enables the simultaneous detection of multiple analytes.

Several studies have reported using SERS for trace detection of ENR with various SERS-active substrates. Tang et al. synthesized SiO_2 membranes embedded with silver nanoparticles, achieving an experimental limit of detection (LOD) of 10 nmol/L (3.58 ppb) for ENR [42]. Another study describes a monolayer silver thin film prepared from self-assembled silver nanoparticles at a water/oil interface for ENR detection, with a calculated detection limit of ENR in the fish matrix of 0.71 μg mL^{-1} (0.71 ppm) [43]. Chen et al. developed Ag/nanocellulose fibers for the in situ detection of ENR, estimating a LOD of 0.069 ppm [44]. Fu and co-authors fabricated a silver nanoparticles-modified microcavity fiber SERS probe to detect ENR in milk, achieving a LOD of 10 μg/mL (10 ppm) [45]. Neng et al. combined SERS and molecularly imprinted polymers (MIPs) to detect ENR, achieving a LOD of 0.25 ng/mL (0.25 ppb) [46]. Zhou et al. developed the petal-like plasmonic nanoparticle clusters-based colloidal SERS method for ENR detection and achieved a LOD of 1.15 μg/kg (1.15 ppb) [47].

In this study, we synthesized AgTNPls using a straightforward, room-temperature, seed-mediated method with minimal reagents. By immobilizing the AgTNPls onto Al-foil, we fabricated a highly sensitive SERS substrate capable of detecting ENR at concentrations as low as 0.39 μg/L (0.39 ppb), a competitive LOD compared to values reported in the literature.

2. Materials and Methods

2.1. Materials

Enrofloxacin (99%) and rhodamine 6G (95%) in powder form were purchased from Sigma-Aldrich. Other chemicals such as silver nitrate ($AgNO_3$) (99.8%), sodium borohydride ($NaBH_4$) (99.8%), trisodium citrate dihydrate ($Na_3C_6H_5O_7{\cdot}2H_2O$) (99%), hydrogen peroxide (H_2O_2) (30 wt.%) were purchased from Xilong Chemical Co., Ltd. (Shantou, China). All chemicals were used as received, without further purification, and deionized (DI) water was used to prepare all solutions.

2.2. Prepararion of the Silver Nanospheres (AgNSs) and Silver Triangular Nanoplate (AgTNPl) Colloids

The synthesis of AgTNPls colloid was performed using a seed-mediated method in two steps at room temperature. In the first step, small AgNSs were synthesized as seeds, and in the second step, these seeds were grown into AgTNPls. Specifically, during seed fabrication, 0.25 mL of 0.01 M $AgNO_3$ solution was added to 24 mL of DI water. Subsequently, 0.7 mL of 0.05 M trisodium citrate (TSC) solution and 0.25 mL of 0.1 M $NaBH_4$ solution were added simultaneously to this mixture with magnetic stirring for 5 min. The solution turns pale yellow (illustrated in Figure 1a), indicating the formation of AgNSs. Immediately after seed preparation, the growth stage was initiated by adding 65 μL of 30% H_2O_2 to the seed solution. The color of the solution gradually changed from pale yellow to reddish yellow, then to green, and finally to blue (illustrated in Figure 1b), indicating the successful formation of AgTNPls. Note that a portion of the AgNS colloid was collected after the seed preparation step to create SERS substrates for comparison with those prepared from AgTNPls.

Figure 1. The colors of the colloid solution during the AgTNPls synthesis process, in which the pale yellow color of the left image (**a**) corresponds to the colloid after the complete addition of $NaBH_4$, and the dark blue color of the right image (**b**) corresponds to the colloid after the addition of H_2O_2.

After synthesis, both AgNSs and AgTNPls colloids were centrifuged at 12,000 rpm. The transparent supernatant was discarded, and the precipitate was retained, washed several times in DI water to remove residual reactants, and then redispersed in DI water.

2.3. Preparations for SERS Measurements

SERS spectra were recorded for rhodamine 6G (R6G) and enrofloxacin (ENR) at low concentrations. For each measurement, 20 μL of either AgNSs or AgTNPls colloid was mixed with 20 μL of R6G or ENR solution and vortexed for 10 s. A 40 μL aliquot of the resultant solution was then dispensed onto a commercial aluminum foil (Al-foil) with a diameter of 2 mm. Before use, the Al-foil was washed with acetone and rinsed with DI water. The sample was dry naturally in air. R6G concentrations ranged from 10^{-5} to 10^{-13} M, while ENR concentrations ranged from 50 to 0.005 mg/L (ppm) (~1.4×10^{-4}–1.4×10^{-8} M).

Analyte solutions at low concentrations were prepared by sequential dilution. Specifically, R6G or ENR powder was dissolved in methanol to prepare stock solution with concentrations of 10^{-2} M or 100 mg/L, respectively. These stock solutions were subsequently diluted with DI water to create R6G in the range of 10^{-5} to 10^{-13} M and ENR solutions in the range of 50 to 0.005 mg/L (ppm).

The main fabrication steps and SERS applications of AgTNPls are schematically illustrated in Figure 2.

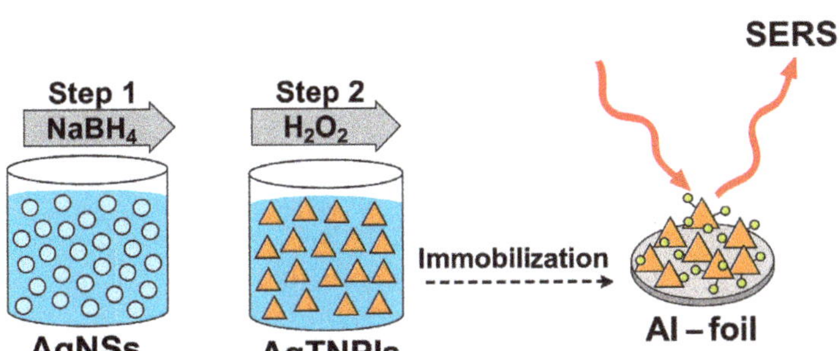

Figure 2. A schematic diagram of the preparation process and SERS application of the AgTNPls.

2.4. Instrumentations

The morphology of AgNSs and AgTNPls was examined using an S-4800 field-emission scanning electron microscope (SEM) (Hitachi, Tokyo, Japan). Size distribution was determined via dynamic light scattering (DLS) using the Zetasizer-Nano ZS (Malvern Instruments, Malvern, UK). Optical absorption spectra of AgNSs and AgTNPls colloids was recorded using UV-Vis absorbance spectroscopy (Shimadzu, Kyoto, Japan) over the range 190–1100 nm. SERS spectra of analytes were measured using portable Raman spectrometer model BWS475-785H with a 785 nm excitation laser (B&W Tek, Newark, DE, USA). The laser spot size was 105 μm with a 20× objective lens magnification. Acquisition time was set to 10 s with 3 accumulations.

3. Results

3.1. The AgTNPls Synthesis Process

C. A. Mirkin and G. S. Metraux [17] pioneered the synthesis of AgTNPls via a thermal synthesis using five reagents: $AgNO_3$, $NaBH_4$, H_2O_2, trisodium citrate (TSC), and polyvinylpyrrolidone (PVP), all mixed in an aqueous solution at room temperature. Subsequent studies have modified this method [9–11,18–22]. To streamline the process, we reduced the number of chemicals, excluding PVP, based on findings suggesting that its omission results in more uniform and stable AgTNPls [19,23].

Mirkin's approach is a seed-mediated method, where small AgNSs (a few to a several tens of nm in diameter) serve as templates for the formation of AgTNPls. However, a key difference in our method is the reversal of the typical addition order of H_2O_2 and $NaBH_4$. In many studies [9–11,17–22], H_2O_2 was added before $NaBH_4$, directly incorporating the seeding step into the formation of AgTNPls. In contrast, we first introduce $NaBH_4$, which reduces $AgNO_3$ to form silver atoms, allowing for AgNSs formation. H_2O_2 is then added to induce the transformation of AgNSs into AgTNPls, with color changes from pale yellow to blue confirming the transition.

Our results indicate that reversing the addition order led to a colloid predominantly composed of AgTNPls, as confirmed by UV-Vis spectroscopy, which displayed the characteristic peaks of AgTNPls (Figure 3, red curve), and further supported by SEM (Figure 4b). By contrast, when H_2O_2 was added before $NaBH_4$, the colloid contained a mixture of AgTNPls and spherical nanoparticles (Figure 4c).

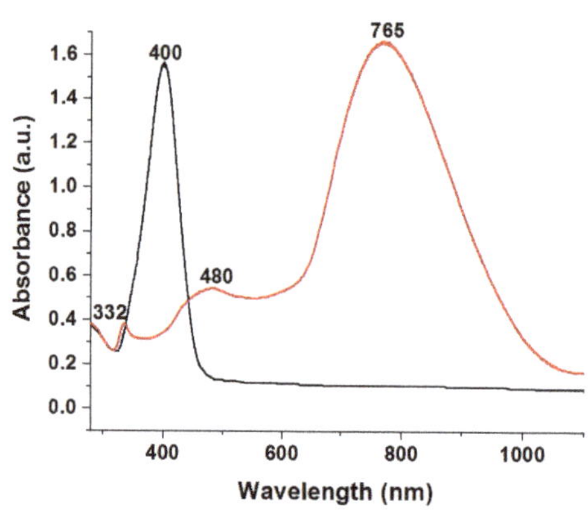

Figure 3. UV-Vis spectra of the silver colloid obtained after the complete addition of $NaBH_4$ (black curve), and after the complete addition of H_2O_2 (red curve).

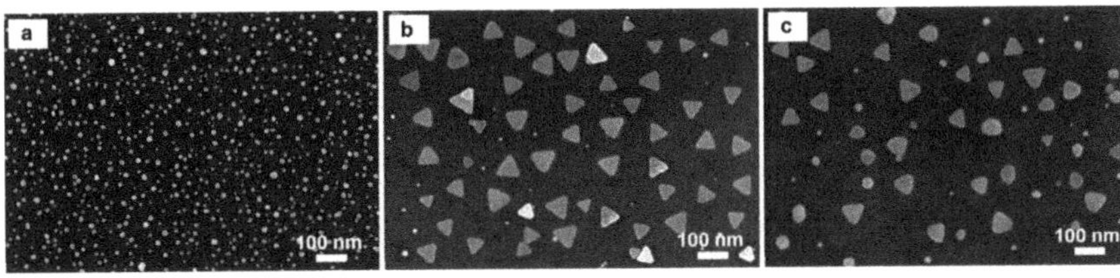

Figure 4. SEM images of the silver colloid precipitate: after the addition of $NaBH_4$ (for seeding) following the procedure of this study (with $NaBH_4$ added before H_2O_2) (**a**); after the addition of H_2O_2 according to the procedure of this study (with $NaBH_4$ added before H_2O_2) (**b**); and after the addition of both H_2O_2 and $NaBH_4$ following the conventional Mirkin procedure (with H_2O_2 added before $NaBH_4$) (**c**).

Our approach clearly separates two distinct steps: the formation of AgNSs as seeds and their transformation into AgTNPls. Specifically, when $NaBH_4$ is added first, it acts as a strong reducing agent that rapidly reduces Ag^+ ions to form small Ag nanospheres (AgNSs) in the 10–25 nm range as seeds. TSC is then added to stabilize these AgNSs and prevent aggregation, which is crucial for achieving uniform AgTNPls. Subsequently, the addition of H_2O_2 (with its dual role as oxidant and reductant) leads to oxidative etching and anisotropic growth of these seeds, favoring triangular nanoplate formation due to surface energy minimization and facet-selective etching. The citrate ions preferentially adsorb on (111) facets, promoting growth in lateral directions rather than vertical ones, which results in the formation of flat, plate-like structures with edge lengths larger than the original seeds [17,19,23–26]. In contrast, adding H_2O_2 before $NaBH_4$ disrupts the controlled seed growth, resulting in a mixture of nanoparticle shapes and less uniformity.

3.2. *Color of Silver Colloids*

The color of the colloidal solution is indicative of the particle shape, as previously established [8,22]. The pale yellow color after the seeding step (Figure 1a) corresponds to spherical AgNPs, while the final blue solution (Figure 1b) results from the formation of AgTNPls. These observations were further confirmed by UV-Vis spectroscopy (Figure 3) and SEM images (Figure 4a,b).

3.3. *UV-Vis Spectroscopy of Silver Colloids*

Figure 3 presents the UV-Vis absorption spectra of the synthesized colloids. The seed colloid exhibited a sharp absorption peak around 400 nm (Figure 3, black curve), typical of small spherical nanoparticles [19,48–51]. In contrast, the final colloid displayed a three-peak absorption pattern (Figure 3, red curve), with peaks at 332 nm, 480 nm and 765 nm, confirming the presence of AgTNPls [17–22]. These peaks correspond to the in-plane dipole and out-of-plane quadrupole resonances [18,20,52,53], confirming the formation of triangular nanoplate structures.

3.4. *SEM Images of Synthesized AgNSs and AgTNPLs*

SEM images (Figure 4) show that AgNSs, formed during the seeding step, are nearly spherical (Figure 4a), with sizes ranging from 10 to 25 nm. After the addition of H_2O_2, the particles transformed into triangular nanoplates (Figure 4b), with side lengths of 40–70 nm. These particles exhibited well-defined shapes and a narrow size distribution, contrasting with the mixture of AgTNPls and spherical particles produced when H_2O_2 was added before $NaBH_4$ (Figure 4c). This comparison highlights the advantages of our method in producing high-quality AgTNPls.

3.5. Size of the Synthesized AgNSs and AgTNPls

Dynamic light scattering (DLS) measurements (Figure 5a,b) revealed that the AgNSs had an average size of 13.5 ± 0.717 nm, with a polydispersity index (PdI) of 0.308, indicating a relatively narrow size distribution. The AgTNPls exhibited an average size of 51.9 ± 0.982 nm and a PdI of 0.452, consistent with the SEM results (Figure 4a,b). The narrow DLS peaks and low PdI values confirm that both AgNSs and AgTNPls are monodispersed [11].

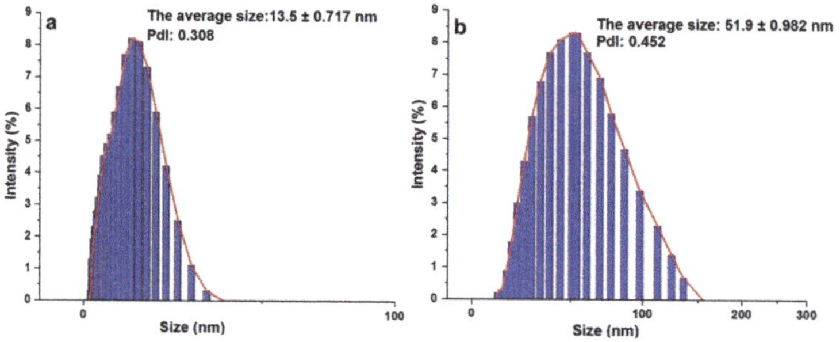

Figure 5. DLS spectra of the colloidal solutions: the pale yellow colloid obtained after the complete addition of $NaBH_4$ to the reaction solution (**a**), and the blue colloid obtained after completing the full synthesis process of AgTNPls according to this study (with $NaBH_4$ added before H_2O_2) (**b**).

The size of AgTNPls can be controlled by adjusting parameters such as the concentration of H_2O_2, the ratio of TSC to Ag^+, and the reaction time. These factors influence the kinetics of oxidative etching and growth. For example, increasing the amount of H_2O_2 can accelerate etching and nucleation, promoting the growth of smaller plates, while decreasing it allows more extended growth, yielding larger plates. This mechanism is consistent with reports of the literature [19,23,24].

3.6. Recording the SERS Spectrum of R6G to Check the Quality of the SERS Substrate

SERS and Raman spectroscopy of R6G, a commonly used organic dye, were conducted to evaluate the quality of the SERS substrates prepared using AgTNPls. Figure 6 displays the SERS spectra of 10^{-5} M R6G recorded with AgTNPls substrates synthesized in this study and coated onto different materials: a microscope glass slide (blue curve), a silicon wafer (red curve), and an Al-foil (black curve). Among these, the substrate with AgTNPls on the glass slide exhibited the lowest enhancement factor, followed by the silicon wafer, while the Al-foil substrate provided the highest enhancement factor.

Further comparison, presented in Figure 7, shows the SERS spectra of 10^{-5} M R6G obtained using AgTNPls synthesized by our method and coated on Al-foil (red curve) versus substrates prepared using alternative methods. The blue curve represents a substrate made with AgTNPls synthesized by a conventional method involving the addition of H_2O_2 prior to $NaBH_4$, while the black curve corresponds to a substrate made with AgNSs (produced during the seed fabrication step). The results indicate that the AgTNPls synthesized via our method achieved a significantly higher SERS enhancement factor due to the production of nearly pure triangular nanoplates. In contrast, the conventional method yielded a mixture of triangular nanoplates and spherical nanoparticles, reducing SERS efficiency. Substrates made from AgNSs, comprising predominantly spherical-shaped particles (Figure 4a), exhibited considerably lower SERS activity, further emphasizing the superior performance of AgTNPls for SERS detection.

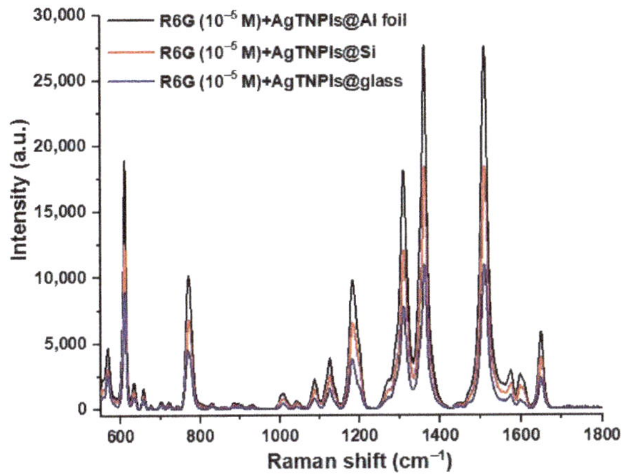

Figure 6. SERS spectrum of 10^{-5} M R6G recorded using an SERS substrate made by coating AgTNPls (synthesized according to the procedure of this study) onto different solid plates: a microscope glass slide (blue curve), a silicon wafer (red curve) and a commercial Al-foil (black curve).

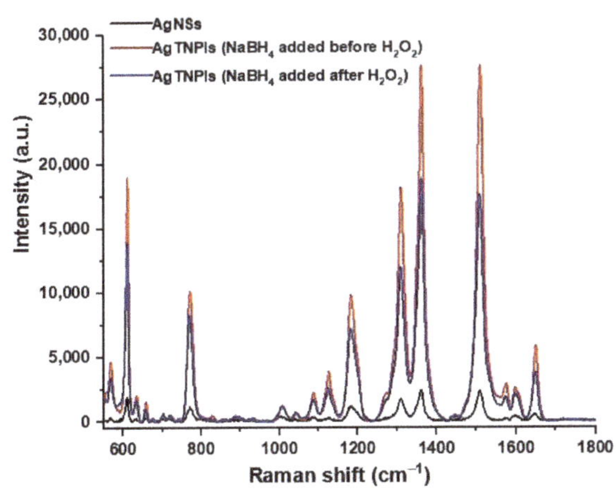

Figure 7. SERS spectrum of 10^{-5} M R6G recorded using an SERS substrate made by coating on a commercial Al-foil with AgNSs obtained after the seed fabrication step (black curve), AgTNPls (and other shaped particles) synthesized by a conventional process in which H_2O_2 was added before adding $NaBH_4$ (blue curve), and AgTNPls synthesized according to the procedure of this study (red curve).

To further demonstrate the benefits of using an Al-foil substrate, SERS spectra of R6G at lower concentrations were recorded with AgTNPls@Al-foil substrates synthesized in this study (Figure 8). The AgTNPls@Al-foil substrate enabled the detection of R6G at concentrations as low as 10^{-13} M, highlighting its remarkable sensitivity for trace-level detection.

The SERS enhancement factor (EF) of the AgTNPls@Al-foil and AgNSs@Al-foil substrates was then evaluated based on the Raman and SERS spectra of R6G. The EF was calculated using the following equation [54,55]:

$$EF = \frac{I_{SERS}C_{RS}}{I_{RS}C_{SERS}}$$

where C_{RS} and C_{SERS} represent the concentrations of R6G on the non-SERS and SERS substrates at which Raman spectra can still be recorded, respectively, and I_{SERS} and I_{RS} are the Raman intensities of a characteristic R6G band under those conditions. The band at 1511 cm^{-1} of R6G was chosen for the calculation due to its sharpness, intensity, and reproducibility, especially at low concentrations.

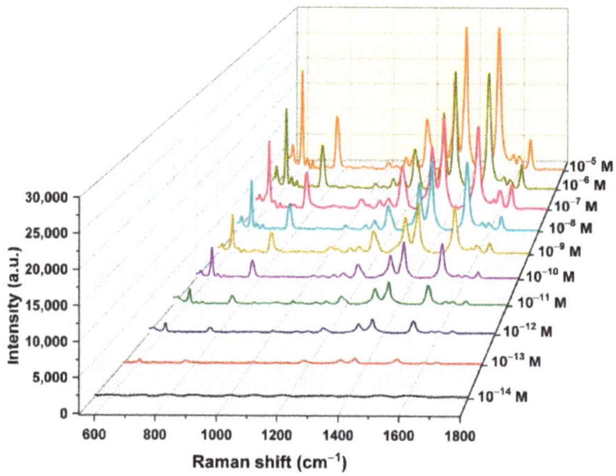

Figure 8. SERS spectra of R6G at different concentrations below 10^{-5} M, recorded using an SERS substrate made by AgTNPls (synthesized according to the procedure of this study) coated on a commercial Al-foil (AgTNPls@Al-foil substrate).

For R6G deposited on a glass plate, C_{RS} was 10^{-3} M, and I_{RS} at 1511 cm^{-1} was approximately 377 in arbitrary unit (a.u.). For the AgTNPls@Al-foil SERS substrate, the lowest detectable R6G concentration was 10^{-13} M, with I_{SERS} at 1511 cm^{-1} being about 120 in a.u., yielding an EF of 3.2×10^9. In comparison, the AgNSs@Al-foil substrate achieved the lowest R6G concentration of 10^{-11} M, with I_{SERS} at 1511 cm^{-1} being about 50 in a.u., resulting in an EF of 1.3×10^7. These findings demonstrate the superior SERS enhancement provided by the AgTNPls@Al-foil substrate.

3.7. Recording the SERS Spectrum of ENR

ENR, a fluoroquinolone antibiotic commonly used in livestock and aquaculture, can accumulate in animal-derived food products, posing a significant risk to food safety and human health due to potential antibiotic resistance. To address the urgent need for effective monitoring methods, we tested our developed AgTNPls@Al-foil SERS substrate's sensitivity towards ENR, specifically targeting concentrations relevant to food safety regulations. The SERS spectra of ENR at concentrations ranging from 50 mg/L to 0.005 mg/L (50–0.005 ppm) were recorded using AgTNPls@Al-foil substrates are presented in Figure 9. The main peaks of ENR, located at 752, 1390, 1476 and 1624 cm^{-1}, can be assigned to the methylene rocking mode, O-C-O symmetric stretching vibration, benzene ring vibration and C=O stretching vibration, respectively [36–41], and remain clearly visible down to a concentration of 0.005 mg/L (ppm). The assignment of observed bands to various vibration modes of ENR is detailed in Table 1.

Such high sensitivity is particularly significant for detecting ENR residues in complex food matrices such as meat, fish, and milk, where residue levels must comply with strict regulatory thresholds (typically in the range of 100–300 μg/kg). The AgTNPls@Al-foil SERS substrate enables rapid, label-free screening of these matrices without the need for complex sample preparation, offering a viable alternative to conventional chromatographic techniques.

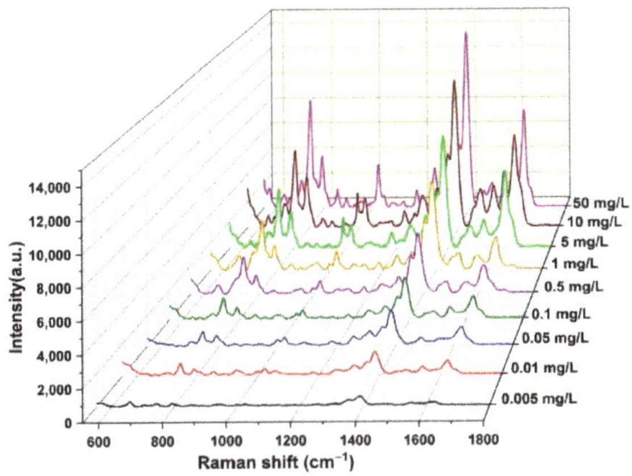

Figure 9. SERS spectra of ENR at concentrations in the range 50–0.005 mg/L (ppm), recorded using an SERS substrate made by AgTNPls (synthesized according to the procedure of this study) coated on a commercial Al-foil (AgTNPls@Al-foil substrate).

Table 1. Characteristic SERS peak assignments of enrofloxacin (ENR) [36–41].

Wavenumber (cm^{-1})	Vibrational Mode Description
752	CH_2 rocking
1390	Symmetric O–C–O stretching (carboxylate group)
1476	Benzene ring breathing
1624	C=O stretching (quinolone ring)

The calibration curve for the 1390 cm^{-1} of ENR (Figure 10) shows a strong linear relationship with ENR concentration in the range of 0.005–50 mg/L, with the regression equation of $y = 88{,}996 \times \text{Log}\, C_{ENR} + 195{,}379$ ($R^2 = 0.9935$). The 1390 cm^{-1} peak was chosen due to its prominence and consistency across varying concentrations of ENR. This peak exhibits high intensity and minimal interference, making it a reliable marker for quantitative analysis. The LOD value for ENR was 0.39 μg/L (ppb) (~1.1×10^{-9} mol/L), estimated using a signal-to-noise ratio of 3 (S/N = 3), demonstrating the excellent sensitivity of the AgTNPls@Al-foil substrate.

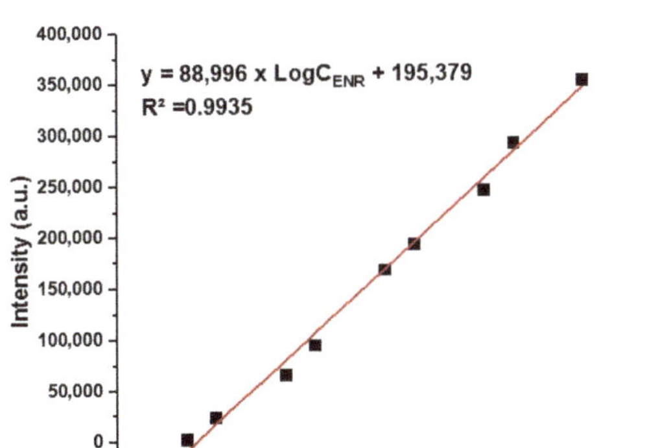

Figure 10. Relationship between the intensity of the ENR 1390 cm^{-1} peak and ENR concentration in the range of 0.005–50 mg/L.

3.8. Uniformity of the SERS Substrate

The uniformity of the AgTNPls@Al-foil SERS substrate was evaluated by recording the SERS spectra at 25 random points on the same substrate of 0.1 mg/L ENR. The relative standard deviation (RSD) of the 1390 cm^{-1} peak intensity was calculated to be 9.83%, shown in Figure 11, indicating excellent uniformity and reproducibility, which is well within the acceptable range (RSD < 20%) for high-quality SERS substrates.

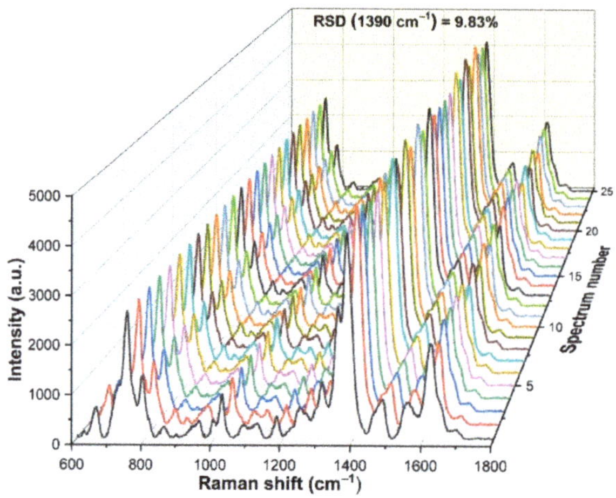

Figure 11. SERS spectra of 0.1 mg/L ENR, recorded from 25 random points on the same AgTNPls@Al-foil substrate.

4. Conclusions

In conclusion, we successfully synthesized a colloid of silver triangular nanoplates (AgTNPls) with an edge length of approximately 52 nm using a straightforward process involving only four common chemicals—silver nitrate, trisodium citrate, sodium borohydride and hydrogen peroxide—at room temperature. By optimizing the sequence of chemical addition, specifically adding hydrogen peroxide after sodium borohydride, we achieved a colloid containing predominantly uniform AgTNPls. The sharp corners of the AgTNPls make them highly effective as SERS substrates, especially when combined with an aluminum foil, which naturally forms a thin layer of aluminum oxide on its surface. This approach enabled the fabrication of an AgTNPls@Al-foil SERS substrate with an enhancement factor of 3.2×10^9. The substrate demonstrated excellent performance in the SERS detection of enrofloxacin (ENR), a commonly used antibiotic in livestock and aquaculture, achieving detection at concentrations as low as 5 µg/L, with a limit of detection of 0.39 µg/L (0.39 ppb). These results highlight the potential application of the developed SERS platform for the detection of antibiotic residues in food matrices, offering a valuable tool for ensuring food safety and supporting regulatory monitoring efforts.

Author Contributions: C.T.A.: investigation, methodology, formal analysis. D.T.C.: conceptualization, methodology, writing—review and editing, supervision, project administration. L.T.-Q.N.: writing—original draft, investigation, data curation, formal analysis. All authors have read and agreed to the published version of the manuscript.

Funding: This work was supported financially by the Institute of Materials Science, Vietnam Academy of Science and Technology (VAST), under project CSCL04.02/23-24.

Data Availability Statement: The data that support the findings of this study are available in this article.

Conflicts of Interest: The authors declare no conflicts of interest.

References

1. Sur, U.K. Surface-enhanced Raman scattering (SERS) spectroscopy: A versatile spectroscopic and analytical technique used in nanoscience and nanotechnology. *Adv. Nano Res.* **2013**, *1*, 111–124. [CrossRef]
2. Fleischmann, M.; Hendra, P.J.; McQuillan, A.J. Raman Spectra of Pyridine Adsorbed at a Silver Electrode. *Chem. Phys. Lett.* **1974**, *26*, 163–166. [CrossRef]
3. Mock, J.J.; Barbic, M.; Smith, D.R.; Schultz, D.A.; Schultz, S. Shape effects in plasmon resonance of individual colloidal silver nanoparticles. *J. Chem. Phys.* **2002**, *116*, 6755–6759. [CrossRef]
4. Cao, D.T.; Ngan, L.T.Q.; Anh, C.T.; Minh, K.N. Fabrication of silver nanostructures in the form of particles, dendrites and flowers on silicon for use in SERS substrates. *Commun. Phys.* **2022**, *32*, 201–212.
5. Schwartzberg, A.M.; Grant, C.D.; Wolcott, A.; Talley, C.E.; Huser, T.R.; Bogomolni, R.; Zhang, J.Z. Unique gold nanoparticle aggregates as a highly active surface-enhanced Raman scattering substrate. *J. Phys. Chem. B* **2004**, *108*, 19191–19197. [CrossRef]
6. Jiang, J.; Shen, Q.; Xue, P.; Qi, H.; Wu, Y.; Teng, Y.; Zhang, Y.; Liu, Y.; Zhao, X.; Liu, X. A highly sensitive and stable SERS sensor for malachite green detection based on Ag nanoparticles in situ generated on 3D MoS2 nanoflowers. *Chem. Sel.* **2020**, *5*, 354–359. [CrossRef]
7. Zavaleta, C.L.; Smith, B.R.; Walton, I.; Doering, W.; Davis, G.; Shojaei, B.; Natan, M.J.; Gambhir, S.S. Multiplexed imaging of surface enhanced Raman scattering nanotags in living mice using noninvasive Raman spectroscopy. *Proc. Natl. Acad. Sci. USA* **2009**, *106*, 13511–13516. [CrossRef]
8. Furletov, A.A.; Apyari, V.V.; Zaytsev, V.D.; Sarkisyan, A.O.; Dmitrienko, S.G. Silver triangular nanoplates: Synthesis and application as an analytical reagent in optical molecular spectroscopy. A review. *TrAC-Trends Anal. Chem.* **2023**, *166*, 117202. [CrossRef]
9. Cheng, D.; He, M.; Ran, J.; Cai, G.; Wu, J.; Wang, X. Depositing a flexible substrate of triangular silver nanoplates onto cotton fabrics for sensitive SERS detection. *Sens. Actuators B Chem.* **2018**, *270*, 508–517. [CrossRef]
10. Xu, R.; Yan, L. Preparation and SERS Study of Triangular Silver Nanoplates. *IOP Conf. Ser. Mater. Sci. Eng.* **2019**, *678*, 012007. [CrossRef]
11. Zannotti, M.; Rossi, A.; Giovannetti, R. SERS activity of silver nanosphere, triangular nanoplates, hexagonal nanoplates and quasi-spherical nanoparticles: Effect of shape and morphology. *Coatings* **2020**, *10*, 288. [CrossRef]
12. Weng, G.; Feng, Y.; Zhao, J.; Li, J.; Zhu, J.; Zhao, J. Size dependent SERS activity of Ag triangular nanoplates on different substrates: Glass vs. paper. *Appl. Sur. Sci.* **2019**, *478*, 275–283. [CrossRef]
13. Kumar, G.; Soni, R.K. Trace-Level Detection of Explosive Molecules with Triangular Silver Nanoplates-Based SERS Substrates. *Plasmonics* **2022**, *17*, 559–573. [CrossRef]
14. Fan, R.; Yang, Y.; Xue, J.; Zhang, F.; Sun, J.; Xu, H.; Jiang, T. Synthesis of Photothermal Stable Triangular Silver Nanoplates for SERS Application and Photokilling of Bacteria. *Chem. Nano Mat.* **2020**, *6*, 148–153.
15. Xue, C.; Métraux, G.S.; Millstone, J.E.; Mirkin, C.A. Mechanistic study of photomediated triangular silver nanoprism growth. *J. Am. Chem. Soc.* **2008**, *130*, 8337–8344. [CrossRef] [PubMed]
16. Dong, X.; Ji, X.; Jing, J.; Li, M.; Li, J.; Yang, W. Synthesis of Triangular Silver Nanoprisms by Stepwise Reduction of Sodium Borohydride and Trisodium Citrate. *J. Phys. Chem. C.* **2010**, *114*, 2070–2074. [CrossRef]
17. Metraux, G.S.; Mirkin, C.A. Rapid thermal synthesis of silver nanoprisms with chemically tailorable thickness. *Adv. Mater.* **2005**, *17*, 412–415. [CrossRef]
18. Torres, V.; Popa, M.; Crespo, D.; Moreno, J.M.C. Silver nanoprism coatings on optical glass substrates. *Microelectron. Eng.* **2007**, *84*, 1665–1668. [CrossRef]
19. Zhang, Q.; Li, N.; Goebl, J.; Lu, Z.; Yin, Y. A systematic study of the synthesis of silver nanoplates: Is citrate a "magic" reagent? *J. Am. Chem. Soc.* **2011**, *133*, 18931–18939. [CrossRef]
20. Pham, V.D.; Chu, H.H.; Le, T.B.; Kasbohm, J. Chemical synthesis and antibacterial activity of novel-shaped silver nanoparticles. *Int. Nano Lett.* **2012**, *2*, 9.
21. Bakar, N.A.; Shapter, J.G.; Salleh, M.M.; Umar, A.A. Self-Assembly of High Density of Triangular Silver Nanoplate Films Promoted by 3-Aminopropyltrimethoxysilane. *Appl. Sci.* **2015**, *5*, 209–221. [CrossRef]
22. Velgosova, O.; Macák, L.; Lisnichuk, M.; Vojtko, M. Synthesis and Analysis of Polymorphic Silver Nanoparticles and Their Incorporation into the Polymer Matrix. *Polymers* **2022**, *14*, 2666. [CrossRef] [PubMed]
23. Li, N.; Zhang, Q.; Quinlivan, S.; Goebl, J.; Gan, Y.; Yin, Y. H_2O_2-Aided Seed-Mediated Synthesis of Silver Nanoplates with Improved Yield and Efficiency. *Chem. Phys. Chem.* **2012**, *13*, 2526–2530. [CrossRef] [PubMed]
24. Parnklang, T.; Lertvachirapaiboon, C.; Pienpinijtham, P.; Wongravee, K.; Thammacharoen, C.; Ekgasit, S. H_2O_2-triggered shape transformation of silver nanospheres to nanoprisms with controllable longitudinal LSPR wavelengths. *RSC Adv.* **2013**, *3*, 12886–12894. [CrossRef]

25. Parnklang, T.; Lamlua, B.; Gatemala, H.; Thammacharoen, C.; Kuimalee, S.; Lohwongwatana, B.; Ekgasit, S. Shape transformation of silver nanospheres to silver nanoplates induced by redox reaction of hydrogen peroxide. *Mater. Chem. Phys.* **2015**, *153*, 127–134. [CrossRef]
26. Musino, D.; Rivard, C.; Novales, B.; Landrot, G.; Capron, I. Tuning of Ag Nanoparticle Properties in Cellulose Nanocrystals/Ag Nanoparticle Hybrid Suspensions by H2O2 Redox Post-Treatment: The Role of the H2O2/AgNP Ratio. *Nanomaterials* **2020**, *10*, 1559. [CrossRef]
27. Hoppmann, E.P.; Yu, W.W.; White, I.M. Inkjet-Printed Fluidic Paper Devices for Chemical and Biological Analytics Using Surface Enhanced Raman spectroscopy. *IEEE J. Sel. Top. Quantum Electron.* **2014**, *20*, 195–204.
28. Pham, T.T.H.; Dien, N.D.; Vu, X.H.; Tran, T.T.; Ca, N.X.; Nguyen, V.T.; Tan, P.M.; Van, H.T.; Phan, V.D. Synthesis and In-Depth Study of the Mechanism of Silver Nanoplate and Nanodecahedra Growth by LED Irradiation for SERS Application. *J. Electron. Mater.* **2020**, *49*, 5009–5027. [CrossRef]
29. Lai, C.-H.; Wang, G.-A.; Ling, T.-K.; Wang, T.-J.; Chiu, P.-K.; Chou Chau, Y.-F.; Huang, C.-C.; Chiang, H.-P. Near infrared surface-enhanced Raman scattering based on starshaped gold/silver nanoparticles and hyperbolic metamaterial. *Sci. Rep.* **2017**, *7*, 5446. [CrossRef]
30. Thu, V.T.; Cuong, N.M.; Cao, D.T.; Hung, L.T.; Ngan, L.T.-Q. Trace detection of ciprofloxacin antibiotic using surface-enhanced Raman scattering coupled with silver nanostars. *Optik* **2022**, *260*, 169043. [CrossRef]
31. Martínez-Torres, P.G.; Martínez-García, M.M.; Cardoso-Ávila, P.E.; Pichardo-Molina, J.L. Facile Nanostructured Substrate Preparation Using Gold Nanocuboids for SERS. *Nanomater. Nanotechnol.* **2015**, *5*, 12. [CrossRef]
32. Liu, X.; Wu, D.; Chang, Q.; Zhou, J.; Zhang, Y.; Wang, Z. Grooved Nanoplate Assembly for Rapidly Surface Enhanced Raman Scattering Detecting. *Nanoscale* **2017**, *9*, 15390–15396. [CrossRef] [PubMed]
33. Gudun, K.; Elemessova, Z.; Khamkhash, L.; Ralchenko, E.; Bukasov, R. Commercial Gold Nanoparticles on Untreated Aluminum Foil: Versatile, Sensitive, and Cost-Effective SERS Substrate. *J. Nanomater.* **2017**, *2017*, 9182025. [CrossRef]
34. Sergiienko, S.; Moor, K.; Gudun, K.; Yelemessova, Z.; Bukasov, R. Nanoparticle–nanoparticle vs. nanoparticle–substrate hot spot contributions to the SERS signal: Studying Raman labelled monomers, dimers and trimers. *Phys. Chem. Chem. Phys.* **2017**, *19*, 4478–4487. [CrossRef]
35. Mukanova, Z.; Gudun, K.; Elemessova, Z.; Khamkhash, L.; Ralchenko, E.; Bukasov, R. Detection of Paracetamol in Water and Urea in Artificial Urine with Gold Nanoparticle@Al Foil Cost-efficient SERS Substrate. *Anal. Sci.* **2018**, *34*, 183–187. [CrossRef] [PubMed]
36. Ha, M.S.; Chung, M.S.; Bae, D.H. Simple Detection of Residual Enrofloxacin in Meat Products Using Microparticles and Biochips. *Food Addit. Contam.-Part A Chem. Anal. Control Expo. Risk Assess.* **2016**, *33*, 817–823. [CrossRef]
37. Aymard, C.; Kanso, H.; Serrano, M.J.; Pagán, R.; Noguer, T.; Istamboulie, G. Development of a New Dual Electrochemical Immunosensor for a Rapid and Sensitive Detection of Enrofloxacin in Meat Samples. *Food Chem.* **2022**, *370*, 131016. [CrossRef]
38. Aslam, B.; Kousar, N.; Javed, I.; Raza, A.; Ali, A.; Khaliq, T.; Muhammad, F.; Khan, J.A. Determination of Enrofloxacin Residues in Commercial Broilers Using High Performance Liquid Chromatography. *Int. J. Food Prop.* **2016**, *19*, 2463–2470. [CrossRef]
39. Fernandez-Torres, R.; Lopez, M.A.B.; Consentino, M.O.; Mochon, M.C. Simultaneous determination of selected veterinary antibiotics and their main metabolites in fish and mussel samples by high-performance liquid chromatography with diode array-fluorescence (HPLC-DAD-FLD) Detection. , *Anal. Lett.* **2011**, *44*, 2357–2372. [CrossRef]
40. Xu, H.; Chen, L.; Sun, L.; Sun, X.; Du, X.; Wang, J.; Wang, T.; Zeng, Q.; Wang, H.; Xu, Y.; et al. Microwave-Assisted Extraction and in Situ Clean-up for the Determination of Fluoroquinolone Antibiotics in Chicken Breast Muscle by LC-MS/MS. *J. Sep. Sci.* **2011**, *34*, 142–149. [CrossRef]
41. Mou, S.A.; Islam, R.; Shoeb, M.; Nahar, N. Determination of chloramphenicol in meat samples using liquid chromatography–tandem mass spectrometry. *Food Sci. Nutr.* **2021**, *9*, 5670–5675.
42. Tang, Y.; Chen, X.; Lv, Y.; Wu, Z.; Chen, F.; Chen, Z. Excellent Surface Enhanced Raman Scattering of SiO2 Fiber Membrane Embedded with Ag Nanoparticles. *J. Inorg. Organomet. Polym. Mater.* **2018**, *28*, 251–257. [CrossRef]
43. Teng, Y.; Wang, Z.; Ren, Z.; Qin, Y.; Pan, Z.; Shao, K.; She, Y.; Huang, W. Interface-Induced Ag Monolayer Film for SurfaceEnhanced Raman Scattering Detection of Water-Insoluble Enrofloxacin. *Plasmonics* **2021**, *16*, 349–358. [CrossRef]
44. Chen, J.; Huang, M.; Kong, L. Flexible Ag/Nanocellulose Fibers SERS Substrate and Its Applications for in-Situ Hazardous Residues Detection on Food. *Appl. Surf. Sci.* **2020**, *533*, 147454. [CrossRef]
45. Fu, X.; Wang, Z.; Li, J.; Ma, S.; Fu, G.; Jin, W.; Bi, W.; Dong, Y. Microcavity Fiber SERS Probe Coated With Ag Nanoparticles For Detecting Antibiotic in Milk. *IEEE Photonics J.* **2021**, *13*, 6800513. [CrossRef]
46. Neng, J.; Wang, Y.; Zhang, Y.; Chen, P.; Yang, K. MIPs–SERS Sensor Based on Ag NPs Film for Selective Detection of Enrofloxacin in Food. *Biosensors* **2023**, *13*, 330. [CrossRef]
47. Zhou, Y.; Wang, H.; Zhao, Z.; Luan, D.; Bian, X.; Lai, K.; Yan, J. Colloidal SERS measurement of enrofloxacin with petaloid nanostructure clusters formed by terminal deoxynucleotidyl transferase catalyzed cytosine-constituted ssDNA. *Food Chem.* **2023**, *429*, 136954. [CrossRef] [PubMed]

48. Velgosova, O.; Macák, L.; Cižmárová, E.; Mára, V. Influence of Reagents on the Synthesis Process and Shape of Silver Nanoparticles. *Materials* **2022**, *15*, 6829. [CrossRef]
49. Lee, P.C.; Meisel, D. Adsorption and surface-enhanced Raman of dyes on silver and gold sols. *J. Phys. Chem.* **1982**, *86*, 3391–3395. [CrossRef]
50. Huang, T.; Xu, X.-H.N. Synthesis and Characterization of Tunable Rainbow Colored Colloidal Silver Nanoparticles Using Single-Nanoparticle Plasmonic Microscopy and Spectroscopy. *Mater. Chem.* **2010**, *20*, 9867–9876. [CrossRef]
51. Zannotti, M.; Vicomandi, V.; Rossi, A.; Minicucci, M.; Ferraro, S.; Petetta, L.; Giovannetti, R. Tuning of hydrogen peroxide etching during the synthesis of silver nanoparticles. An application of triangular nanoplates as plasmon sensors for Hg^{2+} in aqueous solution. *J. Mol. Liq.* **2020**, *309*, 113238. [CrossRef]
52. Xu, X.; Wang, Y.; Wang, H.; Su, H.; Mao, X.; Jiang, L.; Liu, M.; Sun, D.; Hou, S. Synthesis of triangular silver nanoprisms and studies on the interactions with human serum albumin. *J. Mol. Liq.* **2016**, *220*, 14–20. [CrossRef]
53. Wijaya, Y.N.; Kim, J.; Choi, W.M.; Park, S.H.; Kim, M.H. A systematic study of triangular silver nanoplates: One-pot green synthesis, chemical stability, and sensing application. *Nanoscale* **2017**, *9*, 11705–11712. [CrossRef] [PubMed]
54. Zhang, Y.; Teng, Y.; Qin, Y.; Ren, Z.; Wang, Z. Determination of Ciprofloxacin in Fish by Surface-Enhanced Raman Scattering Using a Liquid-Liquid Self-Assembled Gold Nanofilm. *Anal. Lett.* **2020**, *53*, 660–670. [CrossRef]
55. Dies, H.; Siampani, M.; Escobedo, C.; Docoslis, A. Direct Detection of Toxic Contaminants in Minimally Processed Food Products Using Dendritic Surface-Enhanced Raman Scattering Substrates. *Sensors* **2018**, *18*, 2726. [CrossRef]

colloids and interfaces

MDPI

Article

Foaming Properties of *Chlorella sorokiniana* Microalgal Biomass

Despoina Georgiou [1], Aggelos Charisis [1], Athina Theocharidou [1], Christos Ritzoulis [1], Georgia Papapanagiotou [2], Christina Samara [2], Christos Chatzidoukas [2] and Eleni P. Kalogianni [1,*]

[1] Department of Food Science and Technology, International Hellenic University, 57400 Thessaloniki, Greece
[2] Laboratory of Biochemical and Biotechnological Processes (LB2P), Department of Chemical Engineering, Aristotle University of Thessaloniki (AUTH), 54124 Thessaloniki, Greece
* Correspondence: elekalo@ihu.gr; Tel.: +30-2310013907

Abstract: *Chlorella sorokiniana* is a well-studied microalga with significant nutritional potential due to its health-promoting nutrients. *C. sorokiniana* is rich in proteins (~50%), lipids (~14%), vitamins, and other bioactive compounds, making it an attractive ingredient for the food industry. Other properties of *C. sorokiniana*, such as its foaming properties, have not been extensively investigated. The present work examines the foaming properties of *C. sorokiniana* biomass and of its fractions, namely the foaming properties of the whole-cell biomass, the disrupted-cell biomass, the water-soluble protein-rich extract, and the disrupted-cell biomass residue after oil extraction. The water/air interfacial tension, zeta potential, foaming capacity, foam stability, and foam morphology of *C. sorokiniana* biomass and its fractions were determined. Furthermore, the effect of the pH on the foaming properties of the water-soluble protein-rich extract was also investigated. The results show that the examined fractions decrease the water-air interfacial tension and form foams. The type of biomass fraction affects strongly the foam characteristics and foam stability. Furthermore, the stability and characteristics of the foam are significantly affected by pH. Overall, the water-soluble protein-rich extract at pH 7 presented the best foam stability, as the foam remained stable for more than 24 h and had a narrow bubble size distribution. The obtained results suggest that fractionated microalgae biomass could be used as an effective foaming agent in different commercial applications.

Keywords: foam; microalgae; *Chlorella sorokiniana*; foam stability; foaming capacity; interfacial tension

1. Introduction

The current trend in the food industry is to develop natural, environmentally friendly, and health-promoting additives such as emulsifiers, surfactants, and foaming agents to replace existing animal and synthetic additives [1]. Although plant-based solutions made from soybeans and peas already exist, there is a growing interest in utilizing single-celled organisms such as bacteria, microalgae, and fungi [2,3]. Microalgae's high cell growth rates, the low-cost cultivation conditions, and their ability to grow in non-potable waters establish them as a novel and promising source for developing food ingredients and functional food products [4,5].

Furthermore, microalgae are capable of synthesizing an array of biomolecules such as polysaccharides, proteins, nucleic acids, and lipids [6], whereas microalgae proteins have a demonstrably higher nutritional value than conventional protein sources due to the significantly higher overall protein content and the richer composition of essential amino acids [7,8]. Proteins are a major nutrient found in microalgae, with content varying among species and influenced by environmental conditions and nutrient availability [9]. Microalgae exhibit a wide range of protein content, from 6% to 63%. Some microalgae, such as *Arthrospira/Limnospira* ('Spirulina') and Chlorella, are particularly rich in proteins, with some strains up to 70% by weight [10]. Lipids are another major nutrient present in microalgae, with content ranging from 1.9% to 40%. Microalgal lipids are composed of a variety of fatty acids, including mono- and polyunsaturated fatty acids, such as omega-3 and omega-6 fatty acids [10]. Monoraphidium sp. has been found to contain 35.8% alpha-linolenic acid, a higher content than most commonly consumed vegetable oils [11].

The ability of microalgae to produce high nutritive value and surface-active compounds, including proteins, polysaccharides, glycolipids, and phospholipids, makes them a potentially interesting source for various industrial applications such as the interfacial stabilization of emulsions and foams.

It has been found that the whole microalgae biomass or the disrupted biomass cells could be used for the foam formation [12–14] without further fractionation or purification. Several studies showed that the separation into a soluble fraction or the protein isolation may enhance the foaming capacity of microalgae [8,12–15]. Stable foams can be formed from microalgal proteins, such as those derived from *Arthrospira platensis*. *Arthrospira platensis* isolate, obtained through alkaline extraction followed by acidic precipitation, demonstrated superior foaming capability and stability compared to whey protein isolate [9]. Moreover, the protein–concentrate extract exhibited higher foam stability, approximately 80%, than whole *Arthrospira platensis* biomass [8]. Soluble proteins isolated by isoelectric precipitation from *Arthrospira platensis* exhibited pH-dependent foaming capacity, with a significant increase observed at a pH exceeding 7.0 [13,16]. However, these foams lacked stability during storage. Conversely, soluble protein extracts from *Tetraselmis* sp. demonstrated higher foam stability compared to soy proteins, whey proteins, and egg white albumin across a broader pH range of 5.0 to 7.0 [17]. Moreover, acid-hydrolyzed insoluble proteins from *Chlorella protothecoides* showed increased foamability, small bubble diameter, and high stability [18]. Finally, high foam capacity (>95) and high foam stability (97%) for over 180 min of storage have been reported for concentrated proteins by three-phase partitioning from *Chlorella pyrenoidosa* [19].

C. sorokiniana is a green algae with rigid cell walls and a cellular size of several micrometers [20] and shows a remarkable ability to grow in a broad spectrum of environmental conditions such as different temperatures, levels of light intensity, and nutrient concentrations. This microalga has the ability to accumulate lipids (up to 20% or higher), proteins (up to 45% or higher), carbohydrates (up to 38%), and antioxidants, the respective concentrations of which depend on the growth conditions, and has been already used for biofuel production, wastewater treatment, nutraceuticals and functional foods, carbon capture, and utilization [20–24]. Regarding the emulsifying and gelling properties of *C. sorokiniana,* there are currently only a few studies that showed that soluble proteins of *C. sorokiniana* received by cell disruption, centrifugation, and lyophilization had good emulsifying properties at a protein concentration of 1.0 wt% and higher stability at pH $\geq$ 5, whereas the water-soluble protein-rich extract from *C. sorokiniana* could be used for the successful gel formation at a minimum protein concentration of 9.9 g/100 mL at 61 °C [25,26]. Nevertheless, the ability of *C. sorokiniana* to form stable foams has not been examined.

Gas bubble dispersions in liquids that are stabilized by surfactants, proteins, polymers, or particles are known as liquid foams [27]. These foams are intrinsically unstable and typically destabilized within a few hours. Their destabilization arises from a combination of mechanisms: drainage of the liquid due to gravity, bubble coarsening (gas transfer between bubbles driven by pressure differentials), and bubble coalescence (film rupture leading to foam collapse) [27]. Foam stability and its destabilization are affected by various parameters, including surfactant type, surface viscosity, pH, concentration, temperature, and the presence of additional components such as salts, carbohydrates, and lipids [8,28,29].

The scope of this work was to investigate, for the first time to the authors' knowledge, the foaming properties of *C. sorokiniana* microalgal cells produced at semi-pilot scale and their fractions. To this end, whole-cell biomass and three fractions of *C. sorokiniana* biomass were prepared, and their foaming ability and foam stability were investigated. The results are discussed with respect to the properties of the extracts determined in the present study, namely dynamic water/air interfacial tension, surface viscoelasticity, biomass composition, and zeta potential.

2. Materials and Methods

2.1. Microalgal Strain Maintenance, Inoculum Preparation and Semi-Pilot Scale Cultivation

In this study, a strain of the microalgal species C. *sorokiniana* (Figure S1), previously isolated and taxonomically characterized by Psachoulia et al. [30], was utilized. Bold Basal Medium BBM [31] served as the growth medium across all culture scales. Detailed information on the laboratory maintenance of the mother unialgal culture(s) and the procedure followed to generate sufficient inoculum through successive cycles of pre-cultures is provided in the supplementary material. These pre-cultures were conducted in a shaker incubator equipped with cool-white LED strips and an atmospheric air supply system (Figure S2).

Autotrophic Cultivation of C. *sorokiniana* in a Semi-Pilot Circular Pond

The semi-pilot-scale cultivation of C. *sorokiniana* was conducted in an open circular pond with a working volume of 100 L, situated within a laboratory setting (Figure S3). The initial culture density was 120 mg/L of dry cell mass, using an appropriate volume of pre-cultures grown in flasks with the same life history. A detailed description of the semi-pilot cultivation system—including the lighting (Figure S4), aeration (Figure S5), and cleaning/sterilization setups utilizing UV-C lamps (Figure S6)—is provided in the supplementary material. The autotrophic cultivation of C. *sorokiniana* was oriented towards maximizing protein content, following the cultivation strategy proposed by Papapanagiotou et al. [32]. This approach prioritized sufficient nitrogen supply throughout the cultivation period and moderate light intensity to promote protein accumulation. Inorganic carbon supply was confined to suspend lipid and carbohydrate biosynthesis in favor of proteins. Light intensity, measured at the culture's surface, was maintained at 100 μmol m^{-2} s^{-1}, with a 16:8-h light:dark photoperiod. Sodium nitrate doses (250 mg L^{-1}) were added to the culture when the nitrogen level in the medium dropped to 50–60 mg $NaNO_3$ L^{-1}. Continuous agitation and aeration at rates of 30 rpm and 20 L min^{-1}, respectively, were provided, together with daily pH adjustments to the value of 7.0 by adding appropriate microdoses of 5 M HCL solution.

Daily culture monitoring included dry cell weight (DCW) and optical density at 750 nm ($OD_{@750}$) measurements, complemented by frequent nitrogen concentration measurements in the culture to guide nitrogen nitrate supplementation. Water lost through evaporation was replenished daily according to a level indicator with autoclaved sterilized dH_2O. DCW was determined by filtering 5 mL of culture through pre-weighed Whatman 934-AH glass microfiber filters (1.2 μm pore size), followed by rinsing with approximately 40 mL of dH_2O, drying at 45 °C for 24 h, and then weighing using a high-precision microbalance (XP 105, Mettler Toledo, Greifensee, Switzerland). Optical density and remaining nitrogen ($N\text{-}NO_3^-$) concentration measurements were performed in filtrated culture samples using a UV-vis spectrophotometer (DR 3900, HACH, Loveland, CO, USA) and the LCK 339 nitrate cuvette test (HACH), respectively.

Culture samples were also set on a daily observation under an optical microscope (LEICA DM2000, Wetzlar, Germany) for early detection of contamination, given the vulnerability of open microalgae cultivation systems to predators, such as protozoa and rotifers. If contamination was detected, the pH-based treatment described by Georgiou et al. [11] was applied. After 25 days of autotrophic cultivation, the C. *sorokiniana* culture reached a dry cell density of 1 g L^{-1}, at which point the entire 100 L culture was harvested for further processing.

2.2. Biomass Recovery and Fractionation of C. sorokiniana Biomass

At the end of cultivation, the suspension of C. *sorokiniana* culture was firstly concentrated by 95% in volume by a pilot plant centrifugal separator (Westfalia, Model SA 1-02-175) and then further concentrated to a water content of 75% using a centrifuge (Sorvall, Model RC-3, Oelde, Germany) with a rotation speed of 4000 rpm for 20 min at 4 °C

(Figure 1). The concentrated microalgae biomass was used for the preparation of the different biomass fractions (Figure 1).

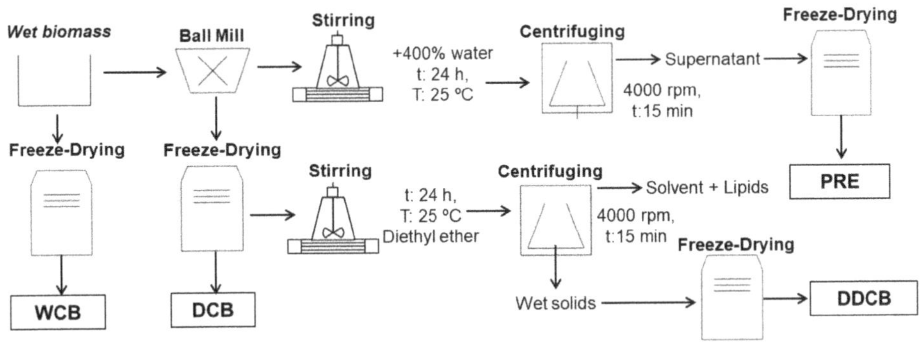

Figure 1. Methods and conditions to recover and fractionate the *C. sorokiniana* biomass into different fractions.

For the whole-cell biomass (WCB), the concentrated biomass was freeze dried for 48 h using Christ Gamma 1–20 freeze dryer (Martin Christ Gefriertrocknungsanlagen GmbH, Osterode am Harz, Germany).

The disrupted-cell biomass (DCB) fraction was prepared using a Fritsch planetary mill (Model pulverisette 5, Fritsch GmbH, Idar-Oberstein, Germany). To disrupt the cells, 45 g of the concentrated microalgae biomass was placed in a grinding bowl of size 250 mL and 400 g zirconium oxide grinding balls of 0.5 mm diameter as added. The experiments were performed at 1500 rpm speed for 60 min. The disrupted biomass was freeze dried for 48 h using a Christ Gamma 1–20 freeze dryer (Martin Christ Gefriertrocknungsanlagen GmbH, Osterode am Harz, Germany).

For the water-soluble protein-rich extract (PRE) fraction, 50 g of the wet disrupted cells were added into a conical flask with 200 mL of water under stirring for 24 h at 25 °C. Next, the suspension was centrifuged at 4000 rpm for 15 min (Rotofix 32A, Hettich, Tuttlingen, Germany). The supernatant was frozen to −18 °C and lyophilized (Martin Christ Gefriertrocknungsanlagen GmbH, Osterode am Harz, Germany) to obtain the dried PB fraction.

Finally, for the preparation of disrupted-cell biomass residue upon oil extraction (DDCB), 10 g of freeze-dried and disrupted microalgal biomass were added into a conical flask with stopper with 100 mL of diethyl ether under stirring for 24 h at 25 °C. At the end of extraction, the solvent solution was centrifuged at 4000 rpm for 15 min (Rotofix 32A, Hettich, Tuttlingen, Germany). The remaining wet solids were frozen to −18 °C and lyophilized (Martin Christ Gefriertrocknungsanlagen GmbH, Osterode am Harz, Germany) to obtain the DB fraction. The effect of the solvent on other compounds, such as proteins, was not investigated.

The freeze-dried fractions were ground using a mortar and pestle to obtain a uniformly powdered form. To prevent moisture absorption, the resulting powder was transferred into airtight glass vials and stored at room temperature until further analysis.

2.3. Biomass Composition

The water content of the microalgae biomass was determined by drying to a constant weight of 5 g of biomass in a porcelain crucible at 105 °C. The measurement was performed in triplicate. The dried sample was used for the ash content determination. For this, the dry samples were heated at 550 °C for 48 h in a furnace. The oil content determination of the biomass was performed after cell disruption. A ball mill (Model pulverisette 5, Fritsch GmbH, Idar-Oberstein, Germany) was applied for cell disruption, and the efficiency of disruption was checked by optical microscopy. The total lipid content was determined by the Bligh and Dyer (BD) extraction method using chloroform and methanol as solvents [33]. The crude protein content was determined by the Kjeldahl method AOAC 954.01 [34]. The conversion factor used was 4.78 based on Lourenco et al. (2004) [35]. The carbohydrate content was determined employing the phenol–sulphuric acid method [36].

2.4. Zeta Potential Measurements

For the zeta potential measurements, a Brookhaven ZetaPALS instrument (Brookhaven Instruments Corporation, Brookhaven, Holtsville, NY, USA) was used. Prior to the measurement, the samples were stirred in ultrapure water at a concentration of 30 mg of dried extract mL^{-1}, subsequently filtered using 1 μm syringe filters, and then measured at 25 °C.

2.5. Water/Air Interfacial Tension and Response of the Interface to a Mechanical Deformation

The pendant drop/axisymmetric drop shape analysis method was applied for the determination of dynamic water/air interfacial tension. For this, a CAM200 (KSV, Biolin Scientific, Stockholm, Sweden) instrument was used, and the One-Attension Software (version 1.8, Biolin Scientific, Västra Frölunda, Sweden) was employed for the analysis of data via the Young–Laplace equation. Measurements were performed at a controlled temperature of 20 °C ± 1 °C. Ultrapure water was used for the solution preparation by adding an equivalent of 2% w/w of each dry fraction. All measurements of water/air interfacial tension were performed over at least 1500 s in triplicate.

The response of the interface to mechanical deformation was performed by using the automatic dispenser of the CAM200 instrument (KSV, Biolin Scientific, Stockholm, Sweden). For this, a drop of volume of 9 μL was formed. After 1800 s (that the system has already reached equilibrium), the interface was dilated at 5–6% compared to its initial interfacial area in a period of 1600 ms (with a rate of 0.25 μL/s), then the area was left constant. The drop images were recorded during the deformation, and the interfacial tension was determined until the interface had reached equilibrium again.

2.6. Size Exclusion Chromatography

Size exclusion chromatograms were obtained by a dedicated device comprising: (i) a SpectraSystem SCM 1000 degasser (Thermo Separation Products, San Jose, CA, USA); (ii) a SpectraSystem P 2000 pump (Thermo Separation Products, San Jose, CA, USA); (iii) a column assembly made of a 2 μm frit (Idex, Oak Harbor, WA, USA), a GPC/SEC PL-Aquagel-OH 50 × 7.5 mm guard column (8 μm) (Varian Inc, Palo Alto, CA, USA), and two GPC/SEC PL-Aquagel-OH 300 × 7.5 mm columns (Varian Inc, Palo Alto, CA, USA), all encased in a Model 605 oven (Scientific Systems Incorporated, State College, PA, USA) thermostated at 30 °C; (iv) a UV detector operating at 280 nm (Rigas Labs, Thessaloniki, Greece) (v) a BI-MwA multi-angle laser light scattering (MALLS) detector (Brookhaven Instruments Corporation, Brookhaven, Holtsville, NY, USA). The results were recorded and treated with ParSec (https://www.brookhaveninstruments.com/product/parsec-gpc-sec/ accessed on 23 September 2024), a dedicated software package (ParSec, Brookhaven Instruments Corporation, Brookhaven, Holtsville, NY, USA). For the measurements, 200 μL of 1.2% w/w of dried sample were injected into an eluent comprising solely ultrapure water at a flow rate of 0.8 mL min^{-1} [37].

2.7. Foam Formation and Characterization

The whole cell biomass and the different fractions of *C. sorokiniana* biomass were dispersed in tap water at a concentration of 10% (w/v). The dispersions were then subjected to a 10 min whipping process using a commercial mixer operating at a speed of 4000 rpm (Moulinex powermix, type HM615, Écully, France). The liquid volume used for foam formation was 40 mL, and the bowl size was 3300 mL.

The pH of the PE dispersions was equal to 7 and was adjusted with acetic acid to 3 and 5 and was equilibrated for 1 h. The pH was measured again after 24 h of storage.

The foaming capacity (FC) was determined using Equation (1):

$$FC\ (\%) = \frac{V_{t0}}{V_{initial}} \cdot 100 \quad (1)$$

where, V_{to} indicates the foam volume after foaming and V_{intial} the liquid volume (mL) used for the foam formation.

The foam stability (FS) of the studied samples was evaluated using Equation (2):

$$FS\ (\%) = \frac{V_t}{V_{t0}} \cdot 100 \quad (2)$$

where V_t is the foam volume remaining after a specific time interval (t).

The foam volume was measured at several time points: immediately after foaming and then at subsequent intervals of 1, 2, 3, 4, and 24 h while the samples rested at room temperature.

2.8. Bubble Size Distribution

To monitor foam structure changes during storage, a 4-h image acquisition process was implemented. A CCD camera (Sony, XC-73CE) connected to a PC captured images of the foam immediately after generation and at subsequent 10-min intervals throughout the test. A scale was included within the image frame to enable subsequent bubble diameter determination using ImageJ 1.54 g software (National Institutes of Health, Bethesda, MD, USA).

3. Results and Discussion

3.1. Biomass Composition

The composition of *C. sorokiniana* biomass is shown in Table 1. Proteins were the predominant component of the microalgae biomass and accounted for 49.2% of the dried biomass, and this finding was in accordance with other studies [21,23,24]. The carbohydrates accounted for 21.8%, which was within the reported range by other studies [21]. Finally, biomass presented a rather low total lipid content, which is lower than those reported in most studies [21,23,24]. It is well established that culture conditions, such as light intensity, temperature, and the cultivation medium composition, influence the composition of microalgal biomass [38,39]. The high protein content of the *C. sorokiniana* biomass could potentially act as a foaming agent [40].

Table 1. Proximate composition of *C. sorokiniana* biomass (results are in dry basis).

	C. sorokiniana
Crude Protein (% *w*/*w*)	49.2 ± 0.2
Total lipids (% *w*/*w*)	13.7 ± 0. 3
Carbohydrates (% *w*/*w*)	21.8 ± 0.1
Ash (% *w*/*w*)	3.5 ± 0.4

3.2. Fractions Characterization

3.2.1. Water/Air Interfacial Tension and Size Exclusion Chromatography Results

The water/air interfacial tension measurements were undertaken to assess the interfacial activity of the obtained fractions and to determine their potential to be used as foaming agents. A rapid decrease in water/air interfacial tension and a low value of interfacial tension indicate rapid and strong adsorption at the interface, which is beneficial for the stabilization of the dispersed air bubbles [41,42]. A rapid decrease in short adsorption times is indicative of small molecular weight surface active compounds. Moreover, since the different fractions are expected to vary in their compositions in carbohydrates, proteins, and lipids the adsorption behavior could indicate which component is mainly adsorbing at the interface. The dynamic water/air interfacial tension of the whole cells and the different fractions of *C. sorokiniana* are presented in Figure 2. The addition of the whole cells and the individual biomass fractions in water decreased the water/air interfacial tension to a different extent. The fraction that was most efficient in decreasing the water/air interfacial tension was the PRE fraction obtained after aqueous extraction. Removing lipids from the DCB fraction (i.e., the DDCB fraction) did not induce a significant change in the water/air interfacial tension in comparison to the DCB fraction as one might expect. Algae contain apolar (intercellular) and polar (mainly at the cell walls) lipids. The exact composition of the lipids and therefore their full removal after diethylether treatment (especially for the polar ones) cannot be known. Therefore, the result is probably due to the presence of surface-active polar lipids (e.g., phospholipids), which are typically present in microalgal cell walls, not having been removed after extraction with diethylether. Although the equilibrium values of water/air interfacial tension for the different fractions were varied, the examined samples showed a rapid decrease in water/air interfacial tension in the first 10 s of the measurement. This indicates rapid adsorption at the interface, which is related to the behavior during foam formation (foamability) [43]. Moreover, the values of the water/air interfacial tension are similar or lower to the values observed by foaming agents [44,45].

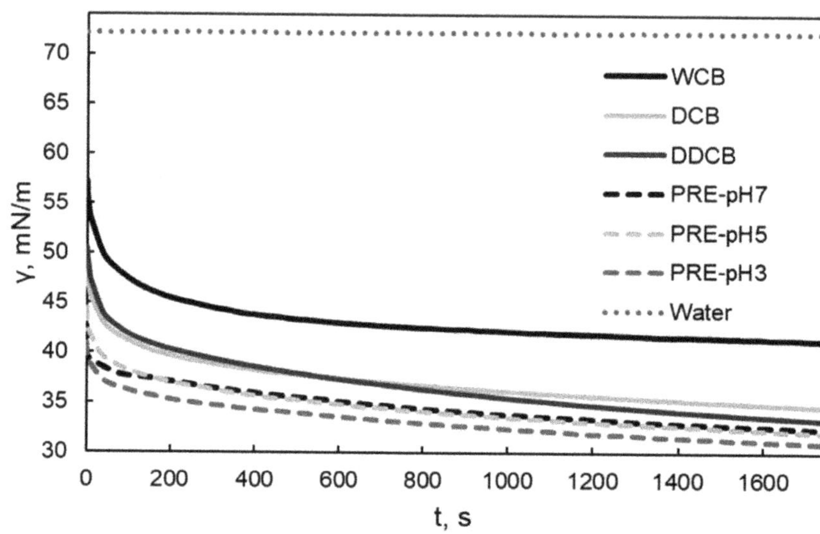

Figure 2. Water/air interfacial tension of the whole cells and the different biomass fractions of *C. sorokiniana* in air. For the definition of WCB, DCB, DDCB, and PRE, refer to Figure 1.

Figure 3 presents the response of water/air interfacial tension to a mechanical deformation. Such measurements are useful to comprehend the mechanical properties of the interface and are therefore related to foam formation and stability. In Figure 3, it can be seen that the interfacial tension abruptly increases upon deformation. The particulate fractions (DCB and DDCB) present a higher relative increase of water/air interfacial tension with respect to the protein extract fraction. The extracted proteins at pH 7 are far from their isoelectric point (see Section 3.2.1), so they are expected to be in the form of a colloidal suspension rather than in particulate form, forming a viscoelastic interfacial layer. On whether the other hand, the particles should form a thicker (due also to size) and more rigid interfacial layer. This is also supported by the rate needed to restore equilibrium, which is faster for the protein extract compared to the particulate fractions. It is interesting that no significant differences are found between the two particulate fractions, although they have different compositions. This is in line with the results in Figure 2, where the water/air interfacial tensions of both fractions are identical and indicate that the removal of apolar lipid species present in the fraction (triacylglycerols) does not affect their adsorption properties.

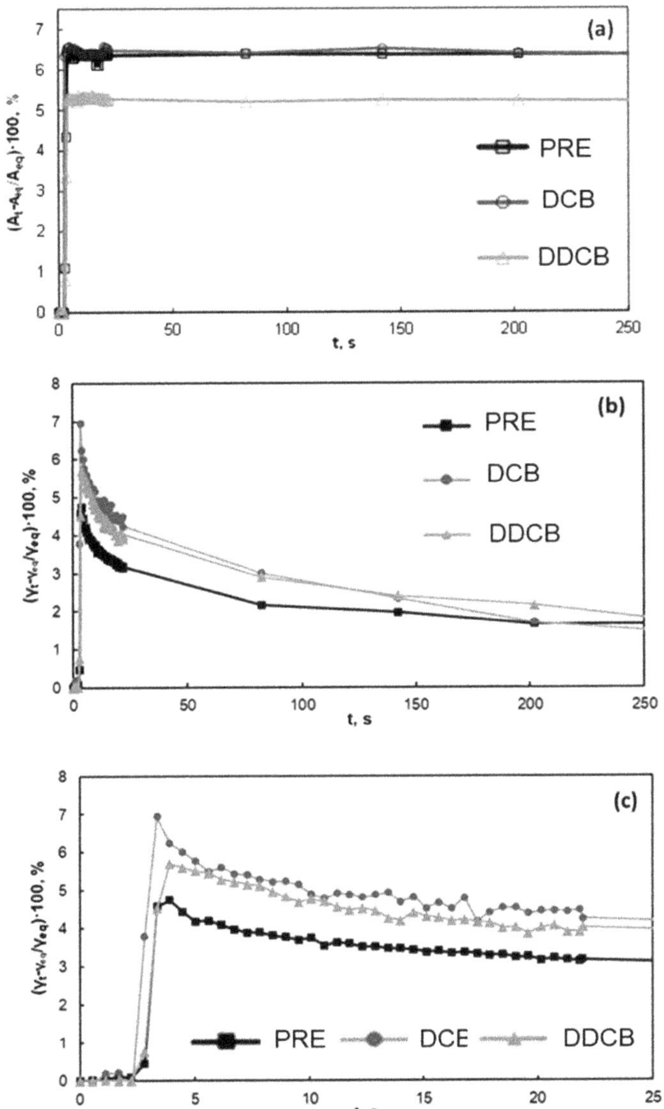

Figure 3. Water/air interfacial tension of the interface of the DCB, DDCB fractions, and PRE fraction at pH 7 following mechanical deformation. (**a**) area changes during deformation, where A_{eq} refers to the area of the drop before deformation and A_t to the area at time t; (**b**,**c**) present the consequent interfacial tension changes, where γ_{eq} refers to the interfacial tension at equilibrium and before deformation and γ_t to the interfacial tension at time t. For the definition of WCB, DCB, DDCB, and PRE, refer to Figure 1.

To better comprehend the nature of the most adsorbing fraction size, exclusion chromatography combined with a MALLS detector was employed. Size exclusion chromatography allows the elution of the larger molecules first, then followed on by molecules of smaller sizes, up to non-polmeric molecules (<100 Da) in the end ("system peak"). The MALLS detector of the present experiment measures light scattering intensity at seven different angles; however only the scattering intensity recorded at 35° is reported, as the scattering was anisotropic, and scattering at small angles provided the strongest scattering. The overall weak scattering suggests open structures in which the refracting index of water dominates, inhibiting the effective diffraction of light. The UV detector was operating at 280 nm as to detect either proteins, due to their relevant absorbance of Tyr and Trp, Phe, and disulfide bonds [46], or other macromolecular compounds containing phenolic entities. In the absence of a strong scattering signal, the determination of the molecular weight was made by comparing the obtained elution times vs. those of standard dextrans; this is a widely used approach, which, however, should be used with caution, as the extent of branching and polydispersity can lead to deviations (e.g., [47]). In this work, it is thus only used to estimate the broad order of size of the eluted populations.

Figure 4 shows a typical size exclusion chromatogram of the PRE-pH7 fraction. It comprises three main macromolecular populations. The largest of all elutes at 4 mL; this is beyond the withholding capacity of the columns, so it can be confidently ascribed to very large supramolecular assemblies of several MDa. These are almost certainly of rather open conformation, as they do not produce strong scattering and are expected to contain phenolics in their structure. A second bimodal population elutes between 8 and 9 mL, corresponding to the elution volumes of dextrans between 2 MDa and 500 kDa, and should be ascribed to polysaccharides. Its dual peaks produce a signal at 35° but less so at 90°, which implies large structures that scatter visible light in an anisotropic fashion. The strong UV absorbance implies the existence of phenolics, most probably aromatic amino acids covalently bound as glycoproteins to the eluting population. A third UV-absorbing dual peak elutes at 13–14 mL, corresponding to the elution time of ~20 kDa dextrans, and most probably comprises proteins. The system peak follows on at larger elution times.

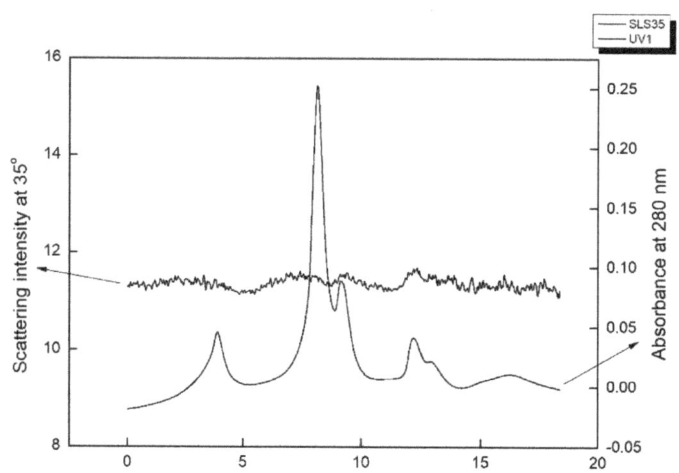

Figure 4. Size exclusion chromatogram of PRE-pH7 fraction: Data are shown for the laser scattering intensity at 35° and for the absorbance at 280 nm.

Functionality across a wide pH range is a crucial requirement for proteins intended for interfacial stabilization in emulsions and foam applications [48]. This is why for this fraction measurements were performed at pH 3, 5, and 7, which are representative of the whole pH range found in food systems (Figure 2). The results show that the changes observed in water/air interfacial tension for the different pH values were rather small. This result was unexpected since commonly the changes in the pH affect strongly the adsorption behavior of proteins, which is expected to be the major (quantitatively) compound in the extract [17]. Nevertheless, besides proteins, which are expected to be the major compound of this extract, the results of interfacial tension indicate the presence of smaller molecular weight biosurfactants. Furthermore, according to the literature, the observed behavior might also be affected by the presence of glycoproteins and polysaccharides, which are common in microalgae extracts and can affect the adsorption behavior [25,49] as well as by the presence of phospholipids transferring from the microalgal cell walls. This was observed for isolated proteins or protein-rich microalgal extracts, where the pH dependence was generally less pronounced compared to the less pure crude extracts or to the whole soluble fraction. However, the interfacial performance of these fractions can be hindered by residual lipids or other species present in the extract [15,48,50].

3.2.2. Zeta Potential

The measurement of zeta potential provides information about the interaction between particles and is therefore useful for the assessment of particle interactions and their potential effect on the formation and stability of foams. The zeta potential was measured for the WBC and for all fractions of *C. sorokiniana* biomass, and the results are presented in Table 2. All fractions besides the PRE fraction present values close to zero, whereas PRE is negatively charged at pH 7 and pH 5.

Table 2. Zeta potential measurements for the WBC and the different fractions of *C. sorokiniana* biomass.

Fraction	Zeta Potential (mV)
WBC	−1.5 (±7.5)
DCB	−5.2 (±7.7)
DDCB	0.5 (±1.9)
PRE-pH7	−12.9 (±1.5)
PRE-pH5	−8.9–(±3.2)
PRE-pH3	−2.4–(±6.8)

3.3. *Foaming Capability and Stability*

Figure 5a shows the FC of the different foams, which were created by the whole cells and the different fractions of *C. sorokiniana* biomass. The PRE fractions, which are rich in proteins and which present the lowest water/air interfacial tension, showed the highest FC, followed by the DCB, DDCB, and WCB with the lowest FC. The effect of pH on the FC of the PE fraction showed that although the PRE-pH7 and PRE-pH3 fractions have similar FC, for the PRE-pH5 fraction, a much higher FC value was observed.

Foam formation involves intense mixing, and this results in rapid renewal of water–air interfaces, intense shear, and at the same time adsorption and liquid film drainage, as well as other phenomena. Therefore, foam formation due to the multiple interfacial phenomena is much more complex than adsorption on the air–water interface, and it is also a more complex situation compared to adsorption at an immobile water–air interface. The observed effect of pH on the FC of foams stabilized by the PRE fraction suggests a potential link between pH and protein adsorption at the air–water interface.

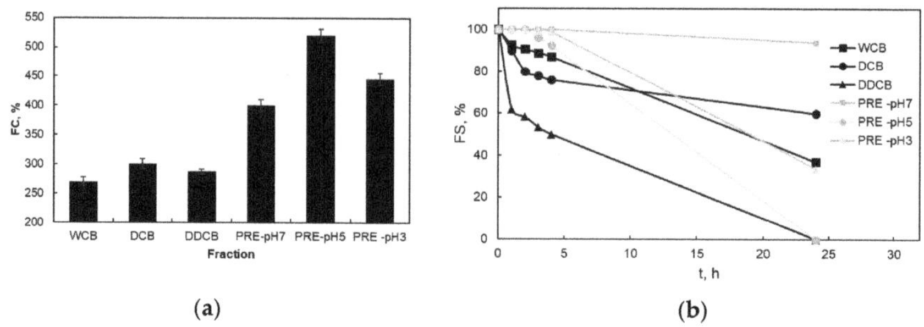

Figure 5. Foaming capacity (FC) (**a**) and foaming stability with time (**b**) of foams prepared by the whole cells and the different fractions of *C. sorokiniana* biomass. For the definition of WCB, DCB, DDCB, and PRE, refer to Figure 1.

Since liquid foams are inherently unstable and usually disappear within a few hours, foam stability is a crucial parameter. The three mechanisms which cause foam destabilization are gravity-driven liquid drainage, bubble coarsening (Ostwald ripening), which is the transfer of gas/liquid from smaller bubbles/droplets to larger ones by diffusion through the continuous phase driven by the difference in Laplace pressure, and bubble coalescence, which is the rupture of films between bubbles causing foam collapse [28,51].

Figure 5b shows the FS of the foams prepared with whole cells and the different fractions of *C. sorokiniana* biomass. PRE-pH7 fraction showed the higher FS, and the foam remained almost stable (94% of the initial volume) after 24 h. Although the water–air interfacial tensions were practically the same for all the pH values examined for RPE, the PRE fractions with pH3 and pH5 presented a different behavior. The volume of the PRE-pH3 foam volume was almost stable for 4 h and then started to decrease. However, the foam volume after 24 h was 33%. On the contrary, the foam volume at pH 5 began to decrease after 2 h and the foam was destroyed completely after 24 h. The enhanced FS observed for the PRE fraction at pH 7 might be attributed to a synergistic effect of increased protein solubility and surface activity of the soluble protein components [16]. Also, the close to zero (given the standard deviation of the measurement) zeta potential of the RPE pH3 and pH5 fractions might facilitate protein aggregate formation with decreased foam stabilizing effects. Nevertheless, with regards to the application of these fractions for instant foams (e.g., in milk-free instant foamed beverages), all the foams presented good stability for two hours, which exceeds by far the time of consumption of such foods.

The foams of the WCB and DCB presented lower FS, however, still acceptable for instant foaming beverages. On the other hand, the foam of the DDCB fraction began to decrease rapidly within 1 h. This was unexpected since the microalga cells contain a relatively high concentration of lipids (Table 1; both apolar lipids such as triacylglycerols and polar ones such as phospholipids), and their removal could be considered beneficial for enhancing foam stability. Apparently, a part of the diethyl ether soluble fraction that was removed presented a stabilizing effect. These findings suggest that the observed bubble stabilization might be attributed to either a particle-stabilized interface (Pickering effect) or a fragment-stabilized interface. Particle-stabilized foams achieve stability through a blocking mechanism, where the particles arrange themselves into dense layers at the interface, hindering film drainage [52,53]. The above findings are also supported by the macroscopic foam properties shown in Figure 6.

Figure 6. Macrostructure of foams developed by the whole cells and different fractions of *C. sorokiniana* biomass immediately after foaming and after 1, 2, and 4 h.

3.4. Foam Morphology and Bubble Size Distribution

Increased foam stability can be related to two key parameters: reduction in foam decay rate and reduction in bubble polydispersity since an increase in polydispersity is related to the Ostwald ripening destabilization mechanism [38,54]. In addition to the evaluation of FC and FS, the foam morphology (Figure 7) and the bubble size distribution over time were investigated (Figure 8).

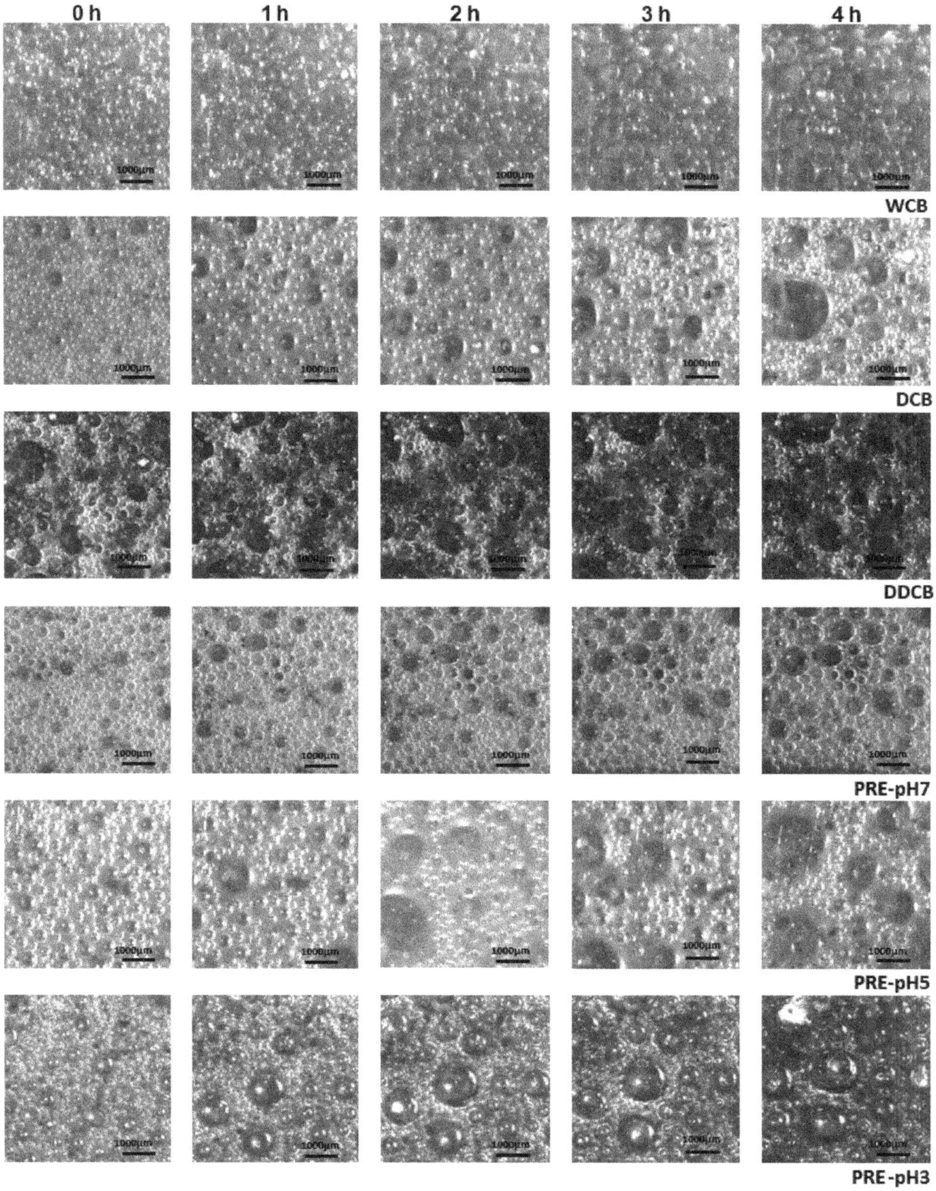

Figure 7. Images of foams made with whole cells and the different fractions of *C. sorokiniana* biomass immediately after foaming and after 1, 2, 3, and 4 h. For the definition of WCB, DCB, DDCB, and PRE, refer to Figure 1. Magnification ×10.

According to the results of Figure 8 at t_0, the box widths of DCB and PRE-pH7 were lower than the other fractions, indicating a narrow bubble size distribution; in contrast, the DDCB fraction showed the largest box width.

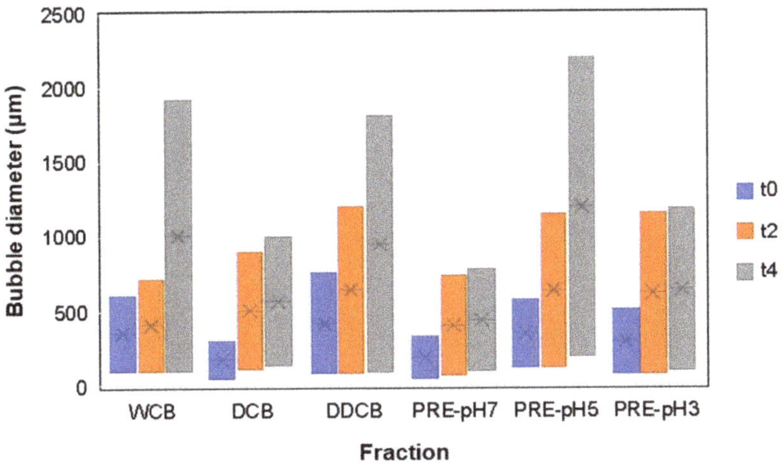

Figure 8. Bubble size distribution of foams made with whole cells and different fractions of *C. sorokiniana* biomass over time. Time points correspond to t_0 immediately after foaming, t_2 after 2 h, and t_4 after 4 h. For the definition of WCB, DCB, DDCB, and PRE, refer to Figure 1.

At t_0, all fractions except DDCB were characterized by a relatively narrow bubble size distribution ranging from 50 to 600 μm. Over the following 4 h, the distribution broadened and became more skewed for all fractions, indicating an increase in polydispersity. Comparing the bubble diameter at t = 4 h, it is clear that the DCB and the DDCB and PRE-pH5 fractions presented an increased polydispersity, and this is an indication of an imminent collapse resulting from Ostwald ripening [55]. This is supported also by the images shown in Figure 8, where the larger foam bubbles grow even larger at the expense of the smaller ones, which decrease in size. Furthermore, in regards to the effect of pH on the performance of the PRE fractions, the evolution of the bubble size distribution over time is in accordance with what is observed for FS, and the reasons for such differences were discussed above.

In addition to the polydispersity observed in bubble size distribution, structural variations were also evident between the foams stabilized by the WCB and the different fractions of *C. sorokiniana* biomass, which is especially visible when comparing the WCB and DCB foams. Whereas Ostwald ripening seems to be a prevailing destabilization mechanism in the DCB foams (at least for the first 4 h of destabilization), WCB foams seem to be untouched by Ostwald ripening but rather by other mechanisms (e.g., drainage). This is supported also by the less spherical shape of the DCB foam bubbles as a function of destabilization time (Figure 7).

Overall, foam made with the PRE-pH7 fraction had the best FS. Furthermore, despite having lower FC than the PRE-pH7 fraction, the foam prepared by WCB and DCB demonstrated good foam stability and morphology. Hence, the different fractions of *C. sorokiniana* showed potential to be alternative foaming agents.

4. Conclusions

This study represents the first investigation, to the authors' knowledge, into the foaming properties of whole-cell *Chlorella sorokiniana* biomass and its various fractions. Measurements of water/air interfacial tension demonstrated that the biomass fractions exhibit surface activity, with interfacial tension values varying based on the specific composition of each fraction. In contrast, variations in the pH of the aqueous microalgal extract had a minimal impact on the water/air interfacial tension. The composition of the fractions also affects the foamability and stability of the observed foams, as well as their morphology and bubble size distribution. The PRE fractions showed the highest FC, followed by the DCB, DDCB fraction, and the WCB. The PRE-pH7 fraction showed the highest FS, and the foam remained almost stable for 24 h. In addition, a satisfactory FS was observed for the WCB and DCB fractions. Investigation over foam destabilization for the first 4 h of foam life revealed different working mechanisms, with Oswald ripening prevailing among the fractions, while the WCB foam destabilization is ruled by other mechanisms. The low FS value of the DDCB foam indicates a pivotal effect of the lipid fraction of microalgal biomass on bubble stabilization. The results of this study highlight the potential of using different fractions of microalgal biomass as foaming agents in various applications, including food products. Especially, the examined fractions, without any modification, (except the DDCB, which is not edible) could be used for instant non-mil foam-forming beverages, taking advantage of the organoleptically pleasant attributes of the foam and the high protein and antioxidant properties. Further investigation prior to commercial application would include studying the effects of fraction concentration, the need for purification of fractions, and potential interactions with other food ingredients

Supplementary Materials: The following supporting information can be downloaded at: https://www.mdpi.com/article/10.3390/colloids8060066/s1, Details on (a) maintenance of the microalga *C. sorokiniana* in the laboratory, (b) production of *C. sorokiniana* inoculum, and (c) semi-pilot cultivation system and preparation steps before inoculation, Figure S1: Microphotograph of *C. sorokiniana* cells. Scale bar 10 μm, Figure S2: Interior view of the shaking incubator used for the production of *C. sorokiniana* inoculum. At the upper inner surface of the incubator, eight dimmable cool-white LED strips (6400 K, 18 W m^{-1}, and 1700 lumen m^{-1}) were attached, while in each flask, filter-sterilized air (pore size 0.2 μm) was also provided. Figure S3: Autotrophic cultivation of *C. sorokiniana* in a semi-pilot circular open pond, Figure S4: Portable artificial lighting panel: (a) the panel positioned above the open pond and (b) a close view of the lighting panel configuration, composed of 30 m of dimmable LED strips (6400 K, 18 W m^{-1}, and 1700 lumen m^{-1}) and three UV-C lamps, Figure S5: Aeration polyethylene tube perimetrically arranged around the pond's bottom; Figure S6: Snapshot of the 24-h UV-C sterilization process of the open pond containing 90 L of distilled water.

Author Contributions: Conceptualization, E.P.K.; methodology, E.P.K.; investigation, D.G., A.C., G.P., C.S., A.T. and C.R.; resources, E.P.K.; writing—original draft preparation, E.P.K., D.G. and G.P.; writing—review and editing, E.P.K. and C.C.; visualization, E.P.K.; supervision, E.P.K.; project administration, E.P.K.; funding acquisition, E.P.K. and C.C. All authors have read and agreed to the published version of the manuscript.

Funding: This research has been co-financed by the European Regional Development Fund of the European Union and Greek national funds through the Operational Program Competitiveness, Entrepreneurship, and Innovation under the call RESEARCH—CREATE—INNOVATE (project code: T2EΔK-02279—project title: "Human nutrition, animal and fish feeding on microalgae derived products through sustainable photosynthetic autotrophic cultures".

Data Availability Statement: The raw data supporting the conclusions of this article will be made available by the authors on request.

Acknowledgments: We thank the Laboratory of Agricultural Constructions and Environmental Control, Department of Agriculture Crop Production and Rural Environment, University of Thessaly, for the provision of the open pond and Verd S.A., for modifications made to the open pond.

Conflicts of Interest: The authors declare no conflicts of interest.

References

1. Chaudhary, A.; Gustafson, D.; Mathys, A. Multi-Indicator Sustainability Assessment of Global Food Systems. *Nat. Commun.* **2018**, *9*, 848. [CrossRef] [PubMed]
2. Caporgno, M.P.; Mathys, A. Trends in Microalgae Incorporation Into Innovative Food Products with Potential Health Benefits. *Front. Nutr.* **2018**, *5*, 58. [CrossRef]
3. Ritala, A.; Häkkinen, S.T.; Toivari, M.; Wiebe, M.G. Single Cell Protein-State-of-the-Art, Industrial Landscape and Patents 2001–2016. *Front. Microbiol.* **2017**, *8*, 2009. [CrossRef] [PubMed]
4. Becker, E.W. Micro-Algae as a Source of Protein. *Biotechnol. Adv.* **2007**, *25*, 207–210. [CrossRef]
5. Radmer', R.J.; Parker, B.C. Commercial Applications of Algae: Opportunities and Constraints. *J. Appl. Phycol.* **1994**, *6*, 93–98. [CrossRef]

6. Xiao, R.; Zheng, Y. Overview of Microalgal Extracellular Polymeric Substances (EPS) and Their Applications. *Biotechnol. Adv.* **2016**, *34*, 1225–1244. [CrossRef] [PubMed]
7. Izydorczyk, M.S.; Biliaderis, C.G. Structural and Functional Aspects of Cereal Arabinoxylans and β-Glucans. *Dev. Food Sci.* **2000**, *41*, 361–384. [CrossRef]
8. Pereira, A.M.; Lisboa, C.R.; Costa, J.A.V. High Protein Ingredients of Microalgal Origin: Obtainment and Functional Properties. *Innov. Food Sci. Emerg. Technol.* **2018**, *47*, 187–194. [CrossRef]
9. Wang, Y.; Tibbetts, S.M.; McGinn, P.J. Microalgae as Sources of High-Quality Protein for Human Food and Protein Supplements. *Foods* **2021**, *10*, 3002. [CrossRef] [PubMed]
10. Koyande, A.K.; Chew, K.W.; Rambabu, K.; Tao, Y.; Chu, D.T.; Show, P.L. Microalgae: A Potential Alternative to Health Supplementation for Humans. *Food Sci. Hum. Wellness* **2019**, *8*, 16–24. [CrossRef]
11. Georgiou, D.; Exarhopoulos, S.; Charisis, A.; Simitsis, S.; Papapanagiotou, G.; Samara, C.; Katsiapi, M.; Kountrias, G.; Bouras, S.; Katsoulas, N.; et al. Valorization of *Monoraphidium* sp. Microalgal Biomass for Human Nutrition Applications. *J. Appl. Phycol.* **2024**, *36*, 1293–1309. [CrossRef]
12. Buchmann, L.; Bertsch, P.; Böcker, L.; Krähenmann, U.; Fischer, P.; Mathys, A. Adsorption Kinetics and Foaming Properties of Soluble Microalgae Fractions at the Air/Water Interface. *Food Hydrocoll.* **2019**, *97*, 105182. [CrossRef]
13. Devi, M.A.; Venkatamaran, L.V. Functional Properties of Protein Products of Mass Cultivated Blue-Green Alga Spirulina Platensia. *J. Food Sci.* **1984**, *49*, 24–27. [CrossRef]
14. Nirmala, C.; Prakash, V.; Venkataraman, L.V. Physico-chemical and Functional Properties of Proteins from Spray Dried Algae (*Spirulina platensis*). *Food/Nahrung* **1992**, *36*, 569–577. [CrossRef]
15. Law, S.Q.K.; Mettu, S.; Ashokkumar, M.; Scales, P.J.; Martin, G.J.O. Emulsifying Properties of Ruptured Microalgae Cells: Barriers to Lipid Extraction or Promising Biosurfactants? *Colloids Surf. B Biointerfaces* **2018**, *170*, 438–446. [CrossRef]
16. Benelhadj, S.; Gharsallaoui, A.; Degraeve, P.; Attia, H.; Ghorbel, D. Effect of PH on the Functional Properties of *Arthrospira (Spirulina) platensis* Protein Isolate. *Food Chem.* **2016**, *194*, 1056–1063. [CrossRef] [PubMed]
17. Schwenzfeier, A.; Lech, F.; Wierenga, P.A.; Eppink, M.H.M.; Gruppen, H. Foam Properties of Algae Soluble Protein Isolate: Effect of PH and Ionic Strength. *Food Hydrocoll.* **2013**, *33*, 111–117. [CrossRef]
18. Dai, L.; Shivananda, R.; Hinrichs, J.; Weiss, J. Foaming of Acid-Hydrolyzed Insoluble Microalgae Proteins from *Chlorella protothecoides*. *Food Biophys.* **2020**, *15*, 368–375. [CrossRef]
19. Waghmare, A.G.; Salve, M.K.; LeBlanc, J.G.; Arya, S.S. Concentration and Characterization of Microalgae Proteins from *Chlorella pyrenoidosa*. *Bioresour. Bioprocess.* **2016**, *3*, 16. [CrossRef]
20. Grossmann, L.; Hinrichs, J.; Weiss, J. Cultivation and Downstream Processing of Microalgae and Cyanobacteria to Generate Protein-Based Technofunctional Food Ingredients. *Crit. Rev. Food Sci. Nutr.* **2020**, *60*, 2961–2989. [CrossRef] [PubMed]
21. Gao, K.; Liu, Q.; Gao, Z.; Xue, C.; Qian, P.; Dong, J.; Gao, Z.; Deng, X. A Dilution Strategy Used to Enhance Nutrient Removal and Biomass Production of *Chlorella sorokiniana* in Frigon Wastewater. *Algal Res.* **2021**, *58*, 102438. [CrossRef]
22. He, Y.; Lian, J.; Wang, L.; Tan, L.; Khan, F.; Li, Y.; Wang, H.; Rebours, C.; Han, D.; Hu, Q. Recovery of Nutrients from Aquaculture Wastewater: Effects of Light Quality on the Growth, Biochemical Composition, and Nutrient Removal of *Chlorella sorokiniana*. *Algal Res.* **2023**, *69*, 102965. [CrossRef]
23. Illman, A.M.; Scragg, A.H.; Shales, S.W. Increase in *Chlorella* strains calorific values when grown in low nitrogen medium. *Enzym. Microb. Technol.* **2000**, *27*, 631–635. [CrossRef]
24. Menegazzo, M.L.; Ulusoy-Erol, H.B.; Hestekin, C.N.; Hestekin, J.A.; Fonseca, G.G. Evaluation of the Yield, Productivity, and Composition of Fatty Acids Methyl Esters (FAME) Obtained from the Lipidic Fractions Extracted from *Chlorella sorokiniana* by Using Ultrasound and Agitation Combined with Solvents. *Biofuels* **2022**, *13*, 519–526. [CrossRef]
25. Ebert, S.; Grossmann, L.; Hinrichs, J.; Weiss, J. Emulsifying Properties of Water-Soluble Proteins Extracted from the Microalgae: *Chlorella sorokiniana* and *Phaeodactylum tricornutum*. *Food Funct.* **2019**, *10*, 754–764. [CrossRef] [PubMed]
26. Grossmann, L.; Hinrichs, J.; Goff, H.D.; Weiss, J. Heat-Induced Gel Formation of a Protein-Rich Extract from the Microalga *Chlorella sorokiniana*. *Innov. Food Sci. Emerg. Technol.* **2019**, *56*, 102176. [CrossRef]
27. Cantat, I.; Cohen-Addad, S.; Elias, F.; Graner, F.; Hohler, R.; Pitois, O.; Rouyer, F.; Saint-Jalmes, F. *Foams: Structure and Dynamics*; Oxford University Press: New York, NY, USA, 2013.
28. Langevin, D. Aqueous Foams: A Field of Investigation at the Frontier between Chemistry and Physics. *ChemPhysChem* **2008**, *9*, 510–522. [CrossRef]
29. Langevin, D. Coalescence in Foams and Emulsions: Similarities and Differences. *Curr. Opin. Colloid Interface Sci.* **2019**, *44*, 23–31. [CrossRef]
30. Psachoulia, P.; Schortsianiti, S.N.; Lortou, U.; Gkelis, S.; Chatzidoukas, C.; Samaras, P. Assessment of Nutrients Recovery Capacity and Biomass Growth of Four Microalgae Species in Anaerobic Digestion Effluent. *Water* **2022**, *14*, 221. [CrossRef]
31. Bold, H.C. Torrey Botanical Society the Morphology of Chlamydomonas Chlamydogama, sp. Nov. *Bull. Torrey Bot. Club* **1949**, *76*, 101–108. [CrossRef]
32. Papapanagiotou, G.; Samara, C.; Chatzidoukas, C. Managing the Interactions of Illumination, Nitrogen, and Sodium Bicarbonate for Autotrophic Biomass Production and Macromolecules Accumulation in the Microalga *Chlorella sorokiniana*. *Algal Res.* **2024**, *80*, 103550. [CrossRef]

33. Bligh, E.G.; Dyer, W.J. A rapid method of total lipid extraction and purification. *Can. J. Biochem. Physiol.* **1959**, *37*, 911–917. [CrossRef] [PubMed]
34. *Official Methods of Analysis*, 17th ed.; AOAC International: Gaithersburg, MD, USA, 2000; p. 954.01.
35. Lourenço, S.O.; Barbarino, E.; Lavín, P.L.; Lanfer Marquez, U.M.; Aidar, E. Distribution of intracellular nitrogen in marine microalgae: Calculation of new nitrogen-to-protein conversion factors. *Eur. J. Phycol.* **2004**, *39*, 17–32. [CrossRef]
36. *Official Methods of Analysis*, 15th ed.; AOAC International: Washington, DC, USA, 1990; p. 988.12.
37. Ritzoulis, C.; Marini, E.; Aslanidou, A.; Georgiadis, N.; Karayannakidis, P.D.; Koukiotis, C.; Filotheou, A.; Lousinian, S.; Tzimpilis, E. Hydrocolloids from Quince Seed: Extraction, Characterization, and Study of Their Emulsifying/Stabilizing Capacity. *Food Hydrocoll.* **2014**, *42*, 178–186. [CrossRef]
38. Frumento, D.; Casazza, A.A.; Al Arni, S.; Converti, A. Cultivation of Chlorella Vulgaris in Tubular Photobioreactors: A Lipid Source for Biodiesel Production. *Biochem. Eng. J.* **2013**, *81*, 120–125. [CrossRef]
39. Breuer, G.; Lamers, P.P.; Martens, D.E.; Draaisma, R.B.; Wijffels, R.H. Effect of Light Intensity, PH, and Temperature on Triacylglycerol (TAG) Accumulation Induced by Nitrogen Starvation in *Scenedesmus obliquus*. *Bioresour. Technol.* **2013**, *143*, 1–9. [CrossRef]
40. Wilson, A.J. *Foams: Physics, Chemistry and Structure*; Springer: Berlin/Heidelberg, Germany, 1989.
41. Beverung, C.J.; Radke, C.J.; Blanch, H.W. Blanch, Protein adsorption at the oil/water interface: Characterization of adsorption kinetics by dynamic interfacial tension measurements. *Biophys. Chem.* **1999**, *81*, 59–80. [CrossRef] [PubMed]
42. Wierenga, P.A.; Meinders, M.B.J.; Egmond, M.R.; Voragen, F.A.G.J.; De Jongh, H.H.J. Protein Exposed Hydrophobicity Reduces the Kinetic Barrier for Adsorption of Ovalbumin to the Air-Water Interface. *Langmuir* **2003**, *19*, 8964–8970. [CrossRef]
43. Denkov, N.; Tcholakova, S.; Politova-Brinkova, N. Physicochemical Control of Foam Properties. *Curr. Opin. Colloid Interface Sci.* **2020**, *50*, 101376. [CrossRef]
44. Bezelgues, J.B.; Serieye, S.; Crosset-Perrotin, L.; Leser, M.E. Interfacial and Foaming Properties of Some Food Grade Low Molecular Weight Surfactants. *Colloids Surf. A Physicochem. Eng. Asp.* **2008**, *331*, 56–62. [CrossRef]
45. Wang, Z.; Zhang, L.; Zhang, X.; Zeng, M.; He, Z.; Chen, J. Interfacial Rheology and Foaming Properties of Soy Protein and Hydrolysates under Acid Condition. *Food Biophys.* **2021**, *16*, 484–491. [CrossRef]
46. Akins, R.E.; Tuan, R.S. Ultrafast Protein Determinations Using Microwave Enhancement. In *The Protein Protocols Handbook*; Humana Press: Totowa, NJ, USA, 2003; pp. 23–30.
47. Gaborieau, M.; Castignolles, P. Size-Exclusion Chromatography (SEC) of Branched Polymers and Polysaccharides. *Anal. Bioanal. Chem.* **2011**, *399*, 1413–1423. [CrossRef] [PubMed]
48. Bertsch, P.; Böcker, L.; Mathys, A.; Fischer, P. Proteins from Microalgae for the Stabilization of Fluid Interfaces, Emulsions, and Foams. *Trends Food Sci. Technol.* **2021**, *108*, 326–342. [CrossRef]
49. Grossmann, L.; Hinrichs, J.; Weiss, J. Solubility and Aggregation Behavior of Protein Fractions from the Heterotrophically Cultivated Microalga *Chlorella protothecoides*. *Food Res. Int.* **2019**, *116*, 283–290. [CrossRef] [PubMed]
50. Böcker, L.; Bertsch, P.; Wenner, D.; Teixeira, S.; Bergfreund, J.; Eder, S.; Fischer, P.; Mathys, A. Effect of *Arthrospira platensis* Microalgae Protein Purification on Emulsification Mechanism and Efficiency. *J. Colloid Interface Sci.* **2021**, *584*, 344–353. [CrossRef]
51. Langevin, D. Recent Advances on Emulsion and Foam Stability. *Langmuir* **2023**, *39*, 3821–3828. [CrossRef]
52. Murray, B.S.; Ettelaie, R. Foam Stability: Proteins and Nanoparticles. *Curr. Opin. Colloid Interface Sci.* **2004**, *9*, 314–320. [CrossRef]
53. Sethumadhavan, G.N.; Nikolov, A.D.; Wasan, D.T. Stability of Liquid Films Containing Monodisperse Colloidal Particles. *J. Colloid Interface Sci.* **2001**, *240*, 105–112. [CrossRef]
54. Marinova, K.G.; Basheva, E.S.; Nenova, B.; Temelska, M.; Mirarefi, A.Y.; Campbell, B.; Ivanov, I.B. Physico-Chemical Factors Controlling the Foamability and Foam Stability of Milk Proteins: Sodium Caseinate and Whey Protein Concentrates. *Food Hydrocoll.* **2009**, *23*, 1864–1876. [CrossRef]
55. Voorhees, P.W. The Theory of Ostwald Ripening. *J. Stat. Phys.* **1985**, *38*, 231–252. [CrossRef]

colloids and interfaces

MDPI

Review

Sodium Alginate: A Green Biopolymer Resource-Based Antimicrobial Edible Coating to Enhance Fruit Shelf-Life: A Review

Anshika Sharma and Arun K. Singh *

Department of Chemistry, M.M. Engineering College, Maharishi Markandeshwar (Deemed to be University), Mullana, Ambala 133207, Haryana, India
* Correspondence: aruniitr09@gmail.com

Abstract: Fruits are a significant source of natural nutrition for human health. However, the perishable nature and short shelf life of fruits lead to spoilage, nutrition safety challenges, and other substantial postharvest losses. Edible coatings have emerged as a novel approach in order to enhance the shelf life of perishable fruits by forming a protective barrier against adverse environmental conditions and microbial infections. Sodium alginate is recognized as an excellent polysaccharide (derived from algae, seaweed, etc.) in the food industry for edible fruit coatings because of its non-allergic, biodegradable, non-toxic (safe for human health), inexpensive, and efficient gel/film-forming properties. However, the hydrophilicity of the polysaccharides is a significant concern to prevent the growth of mold and yeast. In recent years, various plant extracts (containing multiple bioactive compounds, including polyphenolic acids) and nanoparticles have been applied in sodium alginate-based edible films and fruit coatings to enhance antimicrobial activity. This review study summarized recent advancements in fabricating plant extracts incorporating sodium alginate-based films and coatings to enhance fruit shelf life. In addition, approaches to preparing edible films and the basic mechanism behind the role of coating materials in enhancing fruit shelf life are discussed. Moreover, the limitations associated with sodium alginate-based fruit coatings and films have been highlighted.

Keywords: sodium alginate; biopolymer; edible coatings; natural phytochemicals; fruit shelf life

1. Introduction

In a balanced diet, fruits are essential components that facilitate regeneration and cellular detoxification and are also useful for healing several diseases in the human body [1]. Fresh fruits enriched with health-promoting compounds and good appearance are essential features for their marketability and attractiveness to consumers [1,2]. However, the perishable nature of fruits is a significant concern for their short shelf-life after harvest [3]. After the harvesting process, the synthesis of aromatic compounds, generation of reactive oxygen species, ethylene production, and increased respiration rate can degrade the quality of fruits [4]. Apart from the issue of biochemical changes, the contamination of pathogenic microorganisms in postharvest fruits is another serious concern for human health [5].

Traditional approaches to maintaining the quality and extending the shelf-life of fruits are commonly used in waxing, cold storage, modified atmosphere packaging, and chemical preservatives [6]. Waxing is a standard and prevailing method involving applying natural or synthetic waxes to fruits to reduce moisture loss and slow decay. Cold storage is another conventional method of storing fruits at low temperatures to inhibit microbial growth and slow ripening processes. Modified atmosphere packaging alters the gas composition around fruits to create an environment that retards ripening and reduces spoilage. Chemical preservatives have also controlled microbial activity and prolonged fruit freshness. However, these traditional approaches have notable limitations and drawbacks, raising concerns regarding consumer health and environmental impact. This stimulates the exploration and development of modern techniques like edible biodegradable coatings for fruit packaging [7].

Biocompatible materials-based edible coatings/films have emerged as a green environmental approach for increasing shelf-life and protecting fruits from quality deterioration [3,8–11]. Edible films and coatings are typically made from natural biopolymeric edible materials (base matrix) based on polysaccharides (such as starch, chitosan, alginate, pectin, etc.), proteins (such as casein, whey protein, gelatine, or collagen, etc.), and lipids (such as mineral oil, vegetable oil, waxes, natural resins, etc.) or in their combination in optimized proportions [12,13].

Among these biopolymeric materials, alginate is one of the materials that is widely used [14]. Alginate is a polysaccharide derived from seaweeds, which makes it safe for consumption. It is renewable and biocompatible, and the film-forming ability and gelation capacity make it an appropriate material for formulating edible coating [14]. Cations like calcium, sodium, or magnesium bound to alginic acid form an alginate matrix. The biotechnological applications of alginate are based on the alginate molecule and its variations depending on the covalent bonds with cations, such as calcium, sodium, or magnesium. Modifications in the configuration and proportion or ratio of its linear chains of β-D-mannuronic acid and α-L-guluronic acid give rise to its different and unique physicochemical characteristics [15–17]. Because of these structural distinctions, sodium alginate has specific functional qualities that make it an appropriate and edible coating material [13,14]. Due to its unique composition and characteristics, many studies have been conducted on sodium alginate, primarily derived from brown seaweeds like *Laminaria hyperborea* and *Macrocystis pyrifera* [18]. The first stage in the sequential extraction of alginate, which is subsequently purified, is the extraction of alginate-rich seaweed biomass. Sodium alginate has a high G-block content and possesses anionic properties from carboxylic groups attached to guluronic acid residues [19]. Due to its structure, alginic acid can form strong hydrogen bonds with the membranes of microorganisms, which prevents microbial growth and multiplication [19]. Since sodium alginate is hydrophilic, it dissolves readily in water as an edible coating on fruits and vegetables [16,20]. That acts as a barrier against physical and microbiological harm because of their dense gel formation ability [16]. The development of a gel-like viscosity serves as a protective layer against microbiological and physical damage [16]. Its viscoelastic qualities and capacity to form gel allow for even coating on irregular surfaces, which is necessary for efficient preservation [17].

Additionally, the properties of these base matrices of edible coatings can be further improved by incorporating natural phytochemicals to enhance antimicrobial activity, further extending the shelf-life of fruits [11,21]. The phytochemicals, or bioactive substances, present in natural additives like plant extracts, such as clove, contain eugenol [22]; guava leaves contain tannins, phlorotannins, saponins, terpenoids, alkaloids, polyphenols, quercetin, lycopene, and vitamin C [23]; purple onion peel contains anthocyanidin [24], etc., are highly efficient in enhancing the fruit shelf-life. They include significant antioxidant and antibacterial properties that help to maintain the nutritional value, inhibit microbial growth, and delay ripening, thus enhancing fruit shelf-life.

The application involves dipping, spraying, and brushing fruits with a thin layer of the edible coating, creating a barrier that maintains the fruit's moisture content and reduces respiration rates [9,10]. These have tremendous advantages, like controlling respiration (the controlled exchange of gases), inhibiting the action of reactive oxygen species (ROS), and retaining moisture [10]. Therefore, this study summarizes the most commonly used sources of sodium alginate and recent strategies for fabricating alginate-based edible fruit coatings to improve shelf-life. In addition, recent research trends about plant extract-based edible coatings and their importance in the combination of alginate-based coating base matrices are also discussed.

2. Sources and Extraction of Sodium Alginate

These natural sources offer renewable and sustainable alternatives for edible coating formulations and provide benefits such as moisture retention, preservation of sensory qualities in coated food products, and being environmentally friendly substitutes for synthetic polymers [9,19]. Alginate is a biocompatible, biodegradable, and anionic biopolymer with an excellent ability to prepare edible films to preserve food quality and enhance fruit shelf-life [25,26]. Alginates, crucial components (30–60%) of brown algae cell walls, mainly consist of salt combinations of alginic acid with calcium, sodium, potassium, and magnesium [26–28]. Brown seaweeds, particularly those in the *Phaeophyceae* family like *Laminaria* and *Fucoides*, contain high concentrations of alginate, with species such as *Macrocystis pyrifera*, *Ascophyllum nodosum*, and *Laminaria hyperborea* being primary sources globally [27,29,30]. While brown seaweeds are the primary sources of alginates due to their abundance and high alginate content, certain bacteria like *Azotobacter vinelandii* also produce alginate-like polymers, offering alternative sources for extraction [27,30]. As the demand for environmentally friendly food packaging rises, further exploration and refinement of extraction methods from these natural sources promise increased scalability and efficiency in sodium alginate production for edible coating applications [9,13,19]. The extraction process of alginate from brown seaweeds involves several intricate steps to isolate this natural polysaccharide, which has potential applications in edible coatings and various industries [Figure 1]. Initially, the primary sources of brown seaweed are harvested predominantly from cold water regions due to their higher alginic acid content [27]. The seaweeds undergo preparatory steps upon harvesting, including washing, drying, and milling into fine particles. The extraction of alginate relies on converting insoluble alginic acid salts into soluble alginate salts, such as sodium or potassium, through ion exchange reactions [30], wherein the seaweed is treated with a strong alkali following pre-treatment with hydrochloric acid. The alginate isolation process involves several methods to separate alginate from other soluble substances in the crude extract solution, including alcohol precipitation for sodium alginate and calcium chloride solution precipitation for calcium alginate. Mild acid treatments remove undesirable compounds and modify cell wall alginate into alginic acid, maximizing extraction efficacy. After extraction, the alginic acid is recovered in a soluble sodium form through neutralization with sodium carbonate or sodium hydroxide. Insoluble residue is removed through filtration, flotation, or centrifugation, leaving behind soluble alginate [29]. This soluble alginate is then precipitated by conversion into alginic acid or calcium/sodium alginate, with its counter ion adjusted through neutralization with appropriate hydroxides or chlorites. The meticulous extraction process, tailored according to the source and structure of constituents, ensures the efficient extraction of this valuable polysaccharide from brown seaweeds [31].

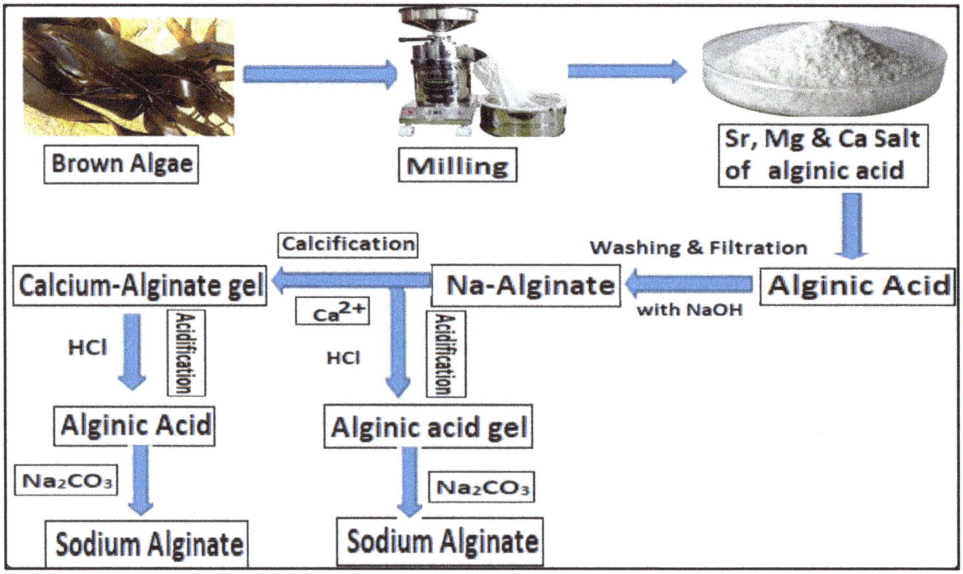

Figure 1. Schematic representation of sodium alginate extraction from brown algae. Reproduced with permission from Ref. [26]. Copyright 2024, Elsevier.

Sodium alginate, obtained through this extraction process, holds immense potential for various applications, particularly in developing edible coatings for fruits, vegetables, and other food products [9,26]. These coatings offer numerous advantages, including shelf-life extension, improved appearance, and maintained nutritional value [13]. Moreover, sodium alginate-based coatings are environmentally friendly and safe for consumption, aligning with the growing demand for sustainable food packaging solutions [32].

2.1. Modification Approaches

When improving the qualities or enabling specific applications, modification techniques for sodium alginate entail changing its chemical structure. These methods in many industrial and biological domains usually seek to enhance properties like solubility, viscosity, stability, or usefulness. Enzymatic, physical (physical blending, physical cross-linking), and chemical modifications (hydroxyl modification, carboxyl modifications, etc.) are frequently used in modification processes [33,34]. Chemical modification involves reacting with substances such as bases, acids, or cross-linking agents to add functional groups to the alginate molecule [33,35]. This may alter some elements, such as the kinetics of drug release or gelation behavior. Physical modification techniques include irradiation, sonication, and mechanical therapy. These procedures can change sodium alginate's molecular weight, particle size, and rheological properties [36]. These alterations usually enhance the material's ability to interact with other materials or function better in specific applications, such as food preservation [33,37].

2.2. Properties of Sodium Alginate

Sodium alginate is a substance extracted from seaweed and widely used in the pharmaceutical and cosmetics industries, food preservation, and preparation of edible films because of their various unique properties (Figure 2) [29]. These are a few of its essential attributes, such as its safety and non-toxic properties, proving its biocompatibility. It forms a transparent film with good mechanical strength and flexibility. It possesses barrier properties against gases and moisture, which help minimize oxidative reactions, maintain the freshness and quality of fruit, and prolong their shelf life during postharvest storage.

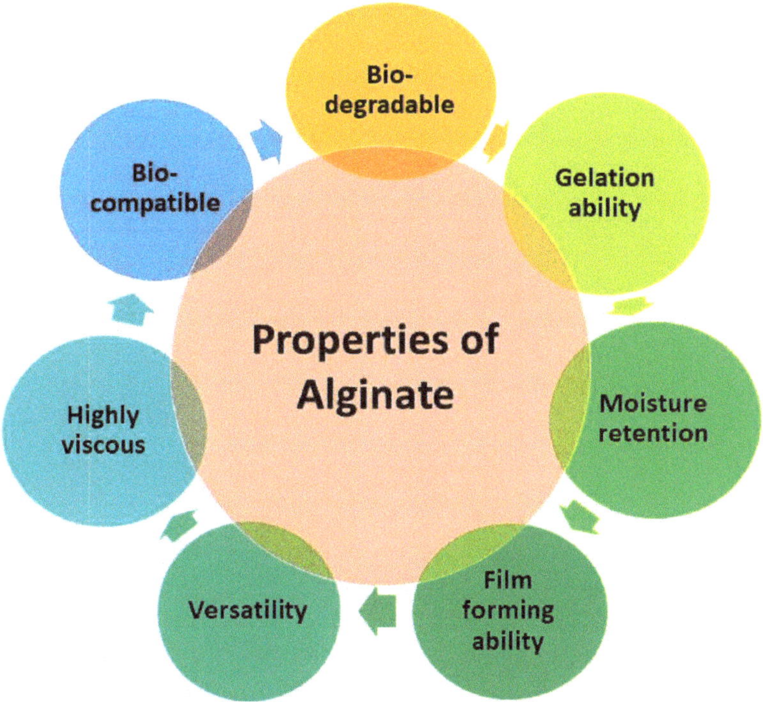

Figure 2. Schematic representation of some common properties of alginate.

Sodium alginate is hydrophilic because the presence of polar groups in large numbers efficiently produces a viscous colloidal solution in water [33,37]. Sodium alginate is insoluble in alcohol, having a content greater than 30% [38]. Due to several cationic or anionic groups in its structure, this polysaccharide exhibits a particular physical property through electrostatic contact [39]. The ability of aqueous solutions of alginate to form firm gels upon the addition of di- and trivalent metal ions, such as bivalent alkaline earth metals (Ca, Sr, and Ba), or trivalent Al and Fe+ ions, is a property that has been widely exploited for the sustained delivery of bioactive molecules. This results from intramolecular bonding and ionic contact between the cations and the carboxylic acid groups on the polymer backbone [40]. The most favored ions for forming microparticles are Ca^{2+} ones, and because Pb^{2+}, Cu^{2+}, and Cd^{2+} are poisonous, their use is restricted [41]. Because of its versatility, sodium alginate can be readily altered for use in various applications by blending it with other polymers, adding natural additives, or combining it with bioactive substances from natural resources, including plant extracts. This ability to adapt enables sodium alginate-based coatings to be customized to the unique requirements of various food products and processing conditions [37].

3. Plant Extract-Mediated Edible Coatings

In recent years, the uses of plant extracts have attracted wide attention in various research areas, including the synthesis of metallic nanoparticles [42–44], biomedical [45,46], antioxidant [47], antibacterial [48], cosmetics [49], sensing [50], edible films, and coating to enhance fruit shelf-life applications [4,51]. In the fabrication of food-grade edible coatings/films, there is an increasing demand for plant extracts for safe and natural additives (antibacterial, antioxidant, etc.) that do not harm human health and the environment. Plant extracts are considered a beneficial resource in formulating edible coatings, especially when preserving fruit's postharvest quality and extending its shelf life. Incorporating plant-based substances and extracts into these coatings enhances their efficacy due to the phytochemicals in plant extracts, such as flavan-3-ols, catechin, gallic acid, quercetin, and kaempferol, which exhibit antimicrobial and antioxidant properties. For example, flavonoids inhibit microbial growth, while antioxidants reduce reactive oxygen species, delaying fruit decay and oxidation and enhancing fruit shelf life. Moreover, it is seen that plant extracts promote the activities of various antioxidant enzymes, contributing to detoxification processes and overall preservation of fruit quality and its nutritional value.

Notably, plant-based coatings extend shelf life and offer nutraceutical benefits, enhancing the nutritional value of treated fruits. Natural antioxidants and antimicrobials are mainly derived from plants through pure bioactive chemicals, fruit, root, leaf, seed extracts, essential oils, peptides, and so on (Figure 3) [52]. Various bioactive substances have significant antimicrobial action, like phenolics, terpenes, aliphatic alcohols, aldehydes, acids, and flavonoids [53]. For instance, extracts from sources like pomegranate peel, garlic, ginger, green tea, and karonda are being explored as potential alternatives to synthetic coatings [54–56]. These extracts, rich in bioactive compounds, mitigate postharvest losses and maintain fruit freshness.

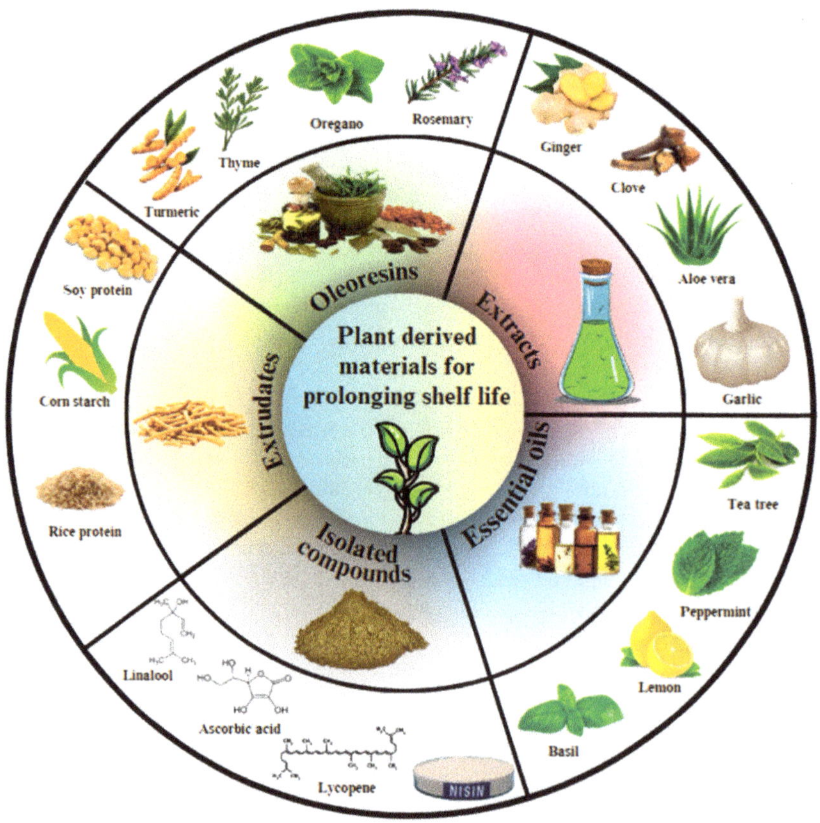

Figure 3. Some commonly used plant-derived resources for the fabrication of edible coatings/films to enhance fruit shelf life. Reproduced with permission from Ref. [52]. Copyright 2024, RSC.

Recent studies have focused on developing edible coatings using plant extracts incorporated with diverse biopolymers (polysaccharides, proteins, lipids, etc.) [4,52]. For instance, cassava starch combined with cinnamon for guava preservation and rice starch blended with carrageenan for plum quality enhancement, peanut shell, pine nutshell, and jujube leaves increase the antioxidant properties of chitosan-based film efficiently [57–59]. The chitosan-based film enriched with dragon fruit extract and blended with ZnO nanoparticles has proven to be an effective way to enhance the films' mechanical strength and thermal stability while reducing water vapor permeability [60]. Incorporation of 0.5% raspberry extract in soya protein isolates films for their enhanced applicability in food packaging as they noticed the improvement in tensile strength and percentage elongation at the break due to raspberry being a rich source of anthocyanin [61]. Pomegranate peel extract combined with mint extract improved the UV light protection of active films made of chitosan and polyvinyl alcohol and their tensile strength [62].

Similarly, it was proposed that adding curry leaf and pomegranate peel extracts to gluten-based films would improve the shelf life of fruits and vegetables, credited to the antimicrobial potential of modified films [63]. Furthermore, it has been reported that extracts from the skin of plums, Italian red grapes, and other elderberry components are rich sources of phenolic compounds, including anthocyanins and flavonoids, which have potent antioxidant properties against various pathogens [64]. In addition to maintaining the physical characteristics, pectin films containing alcoholic extracts of some fruits, such as acerola, cashew apple, and strawberry, demonstrated increased antioxidant capacity. The films with acerola extract exhibited the highest level of antioxidant activity [65]. Likewise, several edible coatings were made by incorporating plant extract into polysaccharides, proteins, or lipids [54,56,66–70]. Plant-based edible coatings offer several advantages compared to synthetic coatings and packaging materials. They are biodegradable, safe for consumption, and contribute to reducing environmental pollution. While a few plant-based coatings have been commercialized, there's immense scope for researchers to explore additional plant extracts for enhancing fruit preservation. This signifies the plant-enriched coatings and their potential to positively impact various physical, biochemical, antioxidant, and enzymatic parameters during the postharvest life of fruit crops. As we switch to safe and sustainable options, plant-based edible coatings emerge as a sustainable and effective solution in postharvest management.

4. Application of Sodium Alginate for Edible Coatings

4.1. Sodium Alginate-Based Edible Coatings Incorporated with Plant Extracts

Sodium alginate has proven to be a potential base material for the development of edible film because of its unique properties, as shown in Figure 2. However, as chemical preservatives, chemical coatings, and chemical pesticides are increasing daily, a safe, sustainable, biodegradable, strong antioxidant and antibacterial coating that is too edible is needed. For this reason, plant extracts have emerged as a remarkable substitute and are used extensively, as these extracts are abundant in bioactive substances like phytochemicals, antioxidants, and antimicrobials. These substances give the films advantages like increased mechanical strength, barrier qualities, antimicrobial and antioxidant activity, and controlled gaseous exchange. As we have already discussed, the importance of phytochemicals present in plant extract is that each component of the extract contributes a unique functionality to the sodium alginate matrix; for example, some contain green tea extract, which offers improved barrier qualities, and extracts of rosemary and oregano, which have strong antioxidant and antibacterial qualities.

Along with these plant extracts, adding plasticizers for better and improved flexibility and reduced brittleness is very important. Water, glycerol, sorbitol, acetylated monoglyceride, polyethylene glycol, sucrose, etc., have been used as plasticizers in developing edible coating. Adding hydrophilic plasticizers to the coating generally promotes water vapor permeability and influences the mechanical properties of the coating material [71]. The fruitful combination of other plant extracts. Some recent studies based on plant extract incorporated with sodium alginate matrix for preparing edible coatings/films to enhance fruit shelf life are summarized in Table 1. The details about coating materials, alginate, and coatings' effect on selected fruits are tabulated.

4.2. Sodium Alginate-Based Edible Coatings Incorporated with Nanoparticles

In addition to using plant extract, researchers have shown significant interest in fabricating sodium alginate-based edible coatings with the incorporation of metallic nanoparticles [72,73]. Incorporating metallic nanoparticles in biopolymeric coatings/films improved the antifungal, antibacterial, and antioxidant properties. Several nanoparticles have been used in alginate-based edible fruit coatings to enhance fruit shelf life. For instance, Khan et al. [72] reported the fabrication of alginate-based (nano-engineered) edible coatings incorporated with titanium dioxide (TiO_2) nanoparticles to extend the shelf life of peach fruits. In this study, initially, TiO_2 nanoparticles (particle size, 5–20 nm) were synthesized via a green synthesis approach using mousami peel extract. The prepared TiO_2 nanoparticles (0.5%) were added to sodium alginate solution (2% *w*/*v*) with glycerol (0.6%) and Tween 80 (1%) to prepare nano-emulsion coatings solution via homogenizer. Thus, the obtained coating nano-emulsion was applied to peach fruits as an active coating suspension to assess their effects on shelf life. After that, coated peach fruit were stored at room temperature and analyzed at intervals of 1, 3, 5, and 7 days. From the results of characterization analysis, it was observed that compared to uncoated, coated peach fruits retained antioxidant activity, reduced percent weight loss, and extended the shelf life of up to 7 days at ambient temperature. Based on experimental observation, authors conclude that the prepared nanoemulsion coating was effective and fruits may be stored at room temperature for a longer time.

Similarly, in another study, Tran et al. [74] reported the fabrication of selenium nanoparticles (Nano Se) incorporated into alginate-based edible coatings and examined their efficacy to extend the shelf life of strawberries at 24 ± 2 °C. In this study, viscous nanoformulations were synthesized using sodium alginate biopolymer (2.0 g) in distilled water (90 mL) and a specific amount (0.62–2.50 mM) of Nao Se (purchased from Merck, Rahway, NJ, USA) with constant stirring of the resultant solution. Strawberry fruits were dipped in the prepared nanoemulsion for about 5 min and then subjected to air drying. The coated and uncoated (control) strawberries were stored at 24 ± 2 °C room temperature for 5 days, and the biochemical and physiological changes were analyzed at definite interval days. The results of the characterization analysis confirmed that 26% less weight was observed in coated fruit samples compared to non-treated (control) samples. In addition, natural appearance was also maintained in coated samples of strawberries, which preserved the highest amount of vitamin C and B9 compared to uncoated references.

Table 1. The combination of different plant extracts along with sodium alginate for edible fruit coatings to enhance shelf-life.

S. No.	Coating Material	Composition	Effect of Coating/Key Findings of the Study	References
1.	Sodium alginate	*Caryophyllus aromaticus* L. clove (flower buds)	Increased the shelf life of fresh-cut apples by up to 14 days with strong antioxidant and antimicrobial activity.	[22]
2.	Sodium alginate	Flowers of the Cananga odorata tree	Effective for maintaining the quality and extending the shelf life of mandarins.	[75]
3.	Sodium alginate	*Pomegranates* were rinsed in sodium hypochlorite solution, and then distilled water, and the fruit peel was manually segregated and oven-dried. Finely sieved powder was left overnight in 80% of 1000 mL ethanol solution at 25 °C.	It has a delayed respiration rate that lowers mass loss, pH, and sensory quality in 'Punjab Beauty' pear fruit, delaying fruit senescence throughout the storage.	[76]
4.	Sodium alginate	Loquat leaf extract and alginate-based coating	Reduce postharvest decay with the preservation of the nutritional quality of Nanfeng tangerines.	[77]
5.	Sodium alginate	Purple onion peel extract and butterfly pea flower extract. The extract was prepared using a solvent–ethanol 60% (*v*/*v*).	Films blended with the two extracts showed more excellent antioxidant, antimicrobial, and UV protection and functioned as a freshness sensor in foods compared to films containing single extracts.	[24]
6.	Alginate and gellan	Sunflower oil, glycerol, and an antibrowning agent—N-acetylcysteine, cross-linking polymer—$CaCl_2$	Extended the shelf life approximately three times compared to uncoated apples, which deteriorate within 4 days of storage. Reduced ethylene production and preserving firmness and color during refrigerated storage of fresh-cut fuji apples.	[78]
7.	Alginate-apple puree	Essential oils (Lemongrass, Oregano oil, Vanillin), glycerol, calcium chloride, N-acetylcysteine,	Coatings with calcium chloride and N-acetylcysteine helped maintain firmness and color. Lemongrass coatings induced severe texture softening. Oregano oil coatings exhibited intense antimicrobial activity against *L. innocua*, effectively inhibiting the growth of psychrophilic aerobic bacteria, yeasts, and molds. Vanillin coatings showed beneficial effects on the shelf life of fresh-cut 'Fuji' apples. Vanillin coatings (0.3% *w*/*w*) were the most effective in terms of sensory quality after 2 weeks of storage.	[79]
8.	Alginate, gelatine	Glycerol, carboxymethylcellulose, sucroesters	2% alginate and 5% gelatine coatings improved appearance, imparting an attractive, natural-looking sheen to the fruit, reduced weight loss, maintained fruit firmness, and preserved freshness. Extended the shelf life of "Bravo de Esmolfe" apples, even at room temperature.	[80]
9.	Alginate	Prebiotics (oligofructose, inulin)	Prebiotic coatings containing oligofructose and inulin remained stable over a 14-day storage period. Coated apple wedges showed a stable browning index, firmness, and acidity compared to the control, with an increase in insoluble solids.	[81]
10.	Alginate	Essential oils	Thyme, cinnamon, and oregano oils were the most potent essential oils against *Listeria monocytogenes*, *Salmonella typhimurium*, and *Staphylococcus aureus*, respectively, with inhibition zones ranging from 12.27 to 24.00 mm. Thyme oil (0.5 $\mu L\ mL^{-1}$) was the most effective in inhibiting microorganisms and respiration, decreasing weight loss, maintaining firmness, and reducing browning reactions in fresh-cut apples.	[82]

Table 1. *Cont.*

S. No.	Coating Material	Composition	Effect of Coating/Key Findings of the Study	References
11.	Sodium Alginate	Glycerol, Lemongrass, sunflower oil	Coated samples with 0.3% and 0.5% (w/v) lemongrass showed significantly lower yeast and mold counts, as well as total plate counts ($p < 0.05$), compared to other samples, extending the shelf life and maintaining the quality of fresh-cut pineapple.	[83]
12.	Sodium Alginate	Thymol, glycerol	Thymol improves sodium alginate film properties, offering antioxidant and antibacterial benefits for fresh-cut apple packaging, potentially substituting polyethylene cling wrap.	[84]
13.	Sodium Alginate	Thyme and Oregano oils	Coatings with 1.0% thyme and oregano oils effectively maintained physicochemical properties and microbiological safety, enhancing papaya's shelf life.	[85]
14.	Sodium Alginate	Cellulose nanocrystals, ginger essential oil.	Enhanced mechanical strength, water vapor resistance, UV–visible light barrier capacity, and antioxidant properties, ultimately improving the postharvest quality of mango fruit.	[86]
15.	Sodium Alginate	Whey protein isolate	Coating effectively slowed down water loss and retarded ripening and browning. It extended the shelf life of bananas by 6 days at 25 °C and exhibited excellent mechanical properties, water resistance, UV shielding, and gas-selective transmission.	[87]
16.	Sodium Alginate	Loquat leaf extract	Enhances antioxidant defense systems, reduces postharvest decay rate and weight loss, and delays senescence in Nanfeng tangerines.	[77]
17.	Sodium Alginate	Pomegranate peel extract	Reduces the respiration rate and mass loss and maintains the firmness and internal quality of 'Punjab Beauty' pears during cold storage.	[76]
18.	Sodium Alginate	Ethanol, glycerol, ylang–ylang oil/cellulose nanocrystals Pickering emulsion	Alginate-based coating containing 0.5% ylang–ylang oil Pickering emulsion (YYO PE) effectively reduces weight loss, maintains firmness, and inhibits the growth of *Penicillium italicum* and *Penicillium digitatum* fungi on mandarins, thereby extending their shelf life.	[75]
19.	Sodium Alginate	Cashew nutshell liquid	Non-toxic film with increased antimicrobial and antioxidant activity and better thermal properties	[88]

5. Conclusions and Future Outlook

Alginate is observed as a remarkable and inexpensive material in active packaging due to its significant properties, like non-toxicity, biodegradability, and gelation capacity, derived from renewable natural sources, thus sustainable and potential to substitute plastic used for the packaging of fruits. The emergence of alginate as an edible coating material plays a vital role in the packaging industry by reducing plastic usage, thus reducing the generation of non-biodegradable waste. Due to its versatility in adjusting towards any additives (plasticizer and cross-linking agent) according to the formulation needed for the development of film/coating, alginate is an appropriate material. Incorporating plant extract into alginate containing various phytochemicals acts as an antioxidant and antimicrobial component, improving and increasing the efficiency of alginate-based coating, thus helping prolong the postharvest shelf life of fruits. Multiple studies have been conducted on incorporating plant extract and nanoparticles into alginate and coating materials. Based on the findings of the literature, it can be concluded that film enriched with plant extracts has the potential to increase tensile strength, reduce brittleness, improve flexibility, and control gas exchange and moisture retention. It helps maintain not only the physical aspects but also the nutritional value of fruit. Though researchers are working with alginate, there are some gaps. However, the focus should be on discovering new formulations and plant extracts that work best with alginate to increase the shelf life of fruits postharvest, thus reducing spoilage in natural environmental conditions at a commercial scale.

Author Contributions: Conceptualization, methodology, formal analysis, investigation, resources, data curation, writing—review and editing, and supervision, A.K.S.; writing—original draft preparation and data analysis, A.S. All authors have read and agreed to the published version of the manuscript.

Funding: The author (A.K.S.) is thankful to the Council of Scientific & Industrial Research (CSIR), New Delhi, India, for the financial support of the research project grant (CSIR/22/0902/23/EMR–II).

Institutional Review Board Statement: Not applicable.

Informed Consent Statement: Not applicable.

Data Availability Statement: Data will be made available on request.

Acknowledgments: The authors acknowledge the support from the Department of Chemistry and Research & Development Cell of Maharishi Markandeshwar (Deemed to be University), Mullana, Ambala, Haryana, India.

Conflicts of Interest: The authors declare no conflicts of interest.

References

1. Nair, M.S.; Tomar, M.; Punia, S.; Kukula-Koch, W.; Kumar, M. Enhancing the functionality of chitosan- and alginate-based active edible coatings/films for the preservation of fruits and vegetables: A review. *Int. J. Biol. Macromol.* **2020**, *164*, 304–320. [CrossRef] [PubMed]
2. Tarancón, P.; Tárrega, A.; González, M.; Besada, C. External quality of mandarins: Influence of fruit appearance characteristics on consumer choice. *Foods* **2021**, *10*, 2188. [CrossRef] [PubMed]
3. Dayeh, N.; Asefpour Vakilian, K.; Azadbakht, M. A fruit edible coating machine to protect the morphological, physiological, and biochemical properties of citrus fruits. *Food Bioprod. Process.* **2024**, *148*, 428–435. [CrossRef]
4. Singh, A.K. Recent advancements in polysaccharides, proteins and lipids based edible coatings to enhance guava fruit shelf-life: A review. *Int. J. Biol. Macromol.* **2024**, *262*, 129826. [CrossRef]
5. Shen, B.; Yan, Z.; Yang, T.; Zhu, L.; Wang, Y.; Jiang, L. Waste citrus pectin/garlic bionanohybrids for edible food preservation. *J. Food Eng.* **2024**, *364*, 111800. [CrossRef]
6. Suhag, R.; Kumar, N.; Petkoska, A.T.; Upadhyay, A. Film formation and deposition methods of edible coating on food products: A review. *Food Res. Int.* **2020**, *136*, 109582. [CrossRef] [PubMed]
7. Chawla, R.; Sivakumar, S.; Kaur, H. Antimicrobial edible films in food packaging: Current scenario and recent nanotechnological advancements—A review. *Carbohydr. Polym. Technol. Appl.* **2021**, *2*, 100024. [CrossRef]
8. Lee, C.R.; Lee, S.J.; Kim, T.I.; Chathuranga, K.; Lee, J.S.; Kim, S.; Kim, M.H.; Park, W.H. Chitosan-gallic acid conjugate edible coating film for perishable fruits. *Food Chem.* **2024**, *463*, 141322. [CrossRef]
9. Manzoor, A.; Yousuf, B.; Pandith, J.A.; Ahmad, S. Plant-derived active substances incorporated as antioxidant, antibacterial or antifungal components in coatings/films for food packaging applications. *Food Biosci.* **2023**, *53*, 102717. [CrossRef]
10. Patil, V.; Shams, R.; Dash, K.K. Techno-functional characteristics, and potential applications of edible coatings: A comprehensive review. *J. Agric. Food Res.* **2023**, *14*, 100886. [CrossRef]
11. Summo, C.; De Angelis, D. The Importance of Edible Films and Coatings for Sustainable Food Development. *Foods* **2022**, *11*, 3221. [CrossRef]
12. Chettri, S.; Sharma, N.; Mohite, A.M. Edible coatings and films for shelf-life extension of fruit and vegetables. *Biomater. Adv.* **2023**, *154*, 213632. [CrossRef] [PubMed]
13. Liu, C.; Jin, T.; Liu, W.; Hao, W.; Yan, L.; Zheng, L. Effects of hydroxyethyl cellulose and sodium alginate edible coating containing asparagus waste extract on postharvest quality of strawberry fruit. *LWT* **2021**, *148*, 111770. [CrossRef]
14. Such, A.; Wisła-Świder, A.; Węsierska, E.; Nowak, E.; Szatkowski, P.; Kopcińska, J.; Koronowicz, A. Edible chitosan-alginate based coatings enriched with turmeric and oregano additives: Formulation, antimicrobial and non-cytotoxic properties. *Food Chem.* **2023**, *426*, 136662. [CrossRef]

15. Hamedi, H.; Kargozari, M.; Shotorbani, P.M.; Mogadam, N.B.; Fahimdanesh, M. A novel bioactive edible coating based on sodium alginate and galbanum gum incorporated with essential oil of Ziziphora persica: The antioxidant and antimicrobial activity, and application in food model. *Food Hydrocoll.* **2017**, *72*, 35–46. [CrossRef]
16. Zhao, N.; Zou, H.; Sun, S.; Yu, C. The interaction between sodium alginate and myofibrillar proteins: The rheological and emulsifying properties of their mixture. *Int. J. Biol. Macromol.* **2020**, *161*, 1545–1551. [CrossRef] [PubMed]
17. Fernández-Escobar, R.; Benlloch, M.; Herrera, E.; García-Novelo, J.M. Effect of traditional and slow-release N fertilizers on growth of olive nursery plants and N losses by leaching. *Sci. Hortic.* **2004**, *101*, 39–49. [CrossRef]
18. Lee, K.Y.; Mooney, D.J. Alginate: Properties and biomedical applications. *Prog. Polym. Sci.* **2012**, *37*, 106–126. [CrossRef] [PubMed]
19. Tavassoli-Kafrani, E.; Shekarchizadeh, H.; Masoudpour-Behabadi, M. Development of edible films and coatings from alginates and carrageenans. *Carbohydr. Polym.* **2016**, *137*, 360–374. [CrossRef]
20. Borchard, W.; Kenning, A.; Kapp, A.; Mayer, C. Phase diagram of the system sodium alginate/water: A model for biofilms. *Int. J. Biol. Macromol.* **2005**, *35*, 247–256. [CrossRef] [PubMed]
21. Guilbert, S.; Gontard, N.; Cuq, B. Technology and applications of edible protective films. *Packag. Technol. Sci.* **1995**, *8*, 339–346. [CrossRef]
22. Pandey, V.K.; Srivastava, S.; Singh, R.; Dar, A.H.; Dash, K.K. Effects of clove essential oil (*Caryophyllus aromaticus* L.) nanoemulsion incorporated edible coating on shelf-life of fresh cut apple pieces. *J. Agric. Food Res.* **2023**, *14*, 100791. [CrossRef]
23. Dinh, T.A.; Le, Y.N.; Pham, N.Q.; Ton-That, P.; Van-Xuan, T.; Ho, T.G.T.; Nguyen, T.; Phuong, H.H.K. Fabrication of antimicrobial edible films from chitosan incorporated with guava leaf extract. *Prog. Org. Coat.* **2023**, *183*, 107772. [CrossRef]
24. Santos, L.G.; Martins, V.G. Multifunctional alginate films blended with polyphenol-rich extract from unconventional edible sources: Bioactive properties, UV-light protection, and food freshness monitoring. *Int. J. Biol. Macromol.* **2024**, *262*, 130001. [CrossRef] [PubMed]
25. Mohammadi, L.; Wardana, A.A.; Tanaka, F.; Tanaka, F. The physicochemical, mechanical, and antifungal properties of sodium alginate film containing Japanese rice vinegar and peppermint (*Mentha piperita*) oil as bio-composite packaging. *Int. J. Biol. Macromol.* **2024**, *281*, 136511. [CrossRef] [PubMed]
26. Kumar, B.; Singh, N.; Kumar, P. A review on sources, modification techniques, properties and potential applications of alginate-based modified polymers. *Eur. Polym. J.* **2024**, *213*, 113078. [CrossRef]
27. Pereira, L.; Cotas, J. Introductory Chapter: Alginates—A General Overview. In *Alginates—Recent Uses of This Natural Polymer*; IntechOpen: London, UK, 2020. [CrossRef]
28. Bojorges, H.; López-Rubio, A.; Martínez-Abad, A.; Fabra, M.J. Overview of alginate extraction processes: Impact on alginate molecular structure and techno-functional properties. *Trends Food Sci. Technol.* **2023**, *140*, 104142. [CrossRef]
29. Shilpa, A.; Agrawal, S.S.; Ray, A.R. Controlled delivery of drugs from alginate matrix. *J. Macromol. Sci.-Polym. Rev.* **2003**, *43*, 187–221. [CrossRef]
30. Mancini, F.; Montanari, L.; Peressini, D.; Fantozzi, P. Influence of alginate concentration and molecular weight on functional properties of mayonnaise. *LWT* **2002**, *35*, 517–525. [CrossRef]
31. Pal, A.; Kamthania, M.C.; Kumar, A. Bioactive Compounds and Properties of Seaweeds—A Review. *Open Access Libr. J.* **2014**, *1*, 1–17. [CrossRef]
32. Yerramathi, B.B.; Muniraj, B.A.; Kola, M.; Konidala, K.K.; Arthala, P.K.; Sharma, T.S.K. Alginate biopolymeric structures: Versatile carriers for bioactive compounds in functional foods and nutraceutical formulations: A review. *Int. J. Biol. Macromol.* **2023**, *253*, 127067. [CrossRef]
33. Yan, P.; Lan, W.; Xie, J. Modification on sodium alginate for food preservation: A review. *Trends Food Sci. Technol.* **2024**, *143*, 104217. [CrossRef]
34. Nezamdoost-Sani, N.; Khaledabad, M.A.; Amiri, S.; Mousavi Khaneghah, A. Alginate and derivatives hydrogels in encapsulation of probiotic bacteria: An updated review. *Food Biosci.* **2023**, *52*, 102433. [CrossRef]
35. Sun, F.; Guo, J.; Liu, Y.; Yu, Y. Preparation, characterizations and properties of sodium alginate grafted acrylonitrile/polyethylene glycol electrospun nanofibers. *Int. J. Biol. Macromol.* **2019**, *137*, 420–425. [CrossRef]
36. Wei, Q.; Zhou, J.; An, Y.; Li, M.; Zhang, J.; Yang, S. Modification, 3D printing process and application of sodium alginate based hydrogels in soft tissue engineering: A review. *Int. J. Biol. Macromol.* **2023**, *232*, 123450. [CrossRef] [PubMed]
37. Metha, C.; Pawar, S.; Suvarna, V. Recent advancements in alginate-based films for active food packaging applications. *Sustain. Food Technol.* **2024**, *2*, 1246–1265. [CrossRef]
38. Das, M.; Senapati, P. Furosemide-loaded alginate microspheres prepared by ionic cross-linking technique: Morphology and release characteristics. *Indian J. Pharm. Sci.* **2008**, *70*, 77–84. [CrossRef] [PubMed]
39. Miyazaki, S.; Kubo, W.; Attwood, D. Oral sustained delivery of theophylline using in-situ gelation of sodium alginate. *J. Control. Release* **2000**, *67*, 275–280. [CrossRef] [PubMed]

40. Katchalsky, A.; Cooper, R.E.; Upadhyay, J.; Wassermann, A. 1028. Counter-ion fixation in alginates. *J. Chem. Soc.* **1961**, 5198–5204. [CrossRef]
41. Agüero, L.; Zaldivar-Silva, D.; Peña, L.; Dias, M. Alginate microparticles as oral colon drug delivery device: A review. *Carbohydr. Polym.* **2017**, *168*, 32–43. [CrossRef] [PubMed]
42. Singh, A.K. A review on plant extract-based route for synthesis of cobalt nanoparticles: Photocatalytic, electrochemical sensing and antibacterial applications. *Curr. Res. Green Sustain. Chem.* **2022**, *5*, 100270. [CrossRef]
43. Singh, A.K.; Bhardwaj, K. Mechanistic understanding of green synthesized cerium nanoparticles for the photocatalytic degradation of dyes and antibiotics from aqueous media and antimicrobial efficacy: A review. *Environ. Res.* **2024**, *246*, 118001. [CrossRef]
44. Bhardwaj, K.; Singh, A.K. Bio-waste and natural resource mediated eco-friendly synthesis of zinc oxide nanoparticles and their photocatalytic application against dyes contaminated water. *Chem. Eng. J. Adv.* **2023**, *16*, 100536. [CrossRef]
45. Kamal, T.; Ul-Islam, M.; Khan, S.B.; Bakhsh, E.M.; Chani, M.T.S. Development of plant extract impregnated bacterial cellulose as a green antimicrobial composite for potential biomedical applications. *Ind. Crops Prod.* **2022**, *187*, 115337. [CrossRef]
46. Li, T.; Sun, W.; Qian, D.; Wang, P.; Liu, X.; He, C.; Chang, T.; Liao, G.; Zhang, J. Plant-derived biomass-based hydrogels for biomedical applications. *Trends Biotechnol.* **2024**. [CrossRef]
47. Kumar, S.; Chauhan, N.; Tyagi, B.; Yadav, P.; Samanta, A.K.; Tyagi, A.K. Exploring bioactive compounds and antioxidant properties of twenty-six Indian medicinal plant extracts: A correlative analysis for potential therapeutic insights. *Food Humanit.* **2023**, *1*, 1670–1679. [CrossRef]
48. Singh, A.K. Flower extract-mediated green synthesis of bimetallic Cu–Zn oxide nanoparticles and its antimicrobial efficacy in hydrocolloid films. *Bioresour. Technol. Rep.* **2022**, *18*, 101034. [CrossRef]
49. Xie, M.; Jiang, Z.; Lin, X.; Wei, X. Application of plant extracts cosmetics in the field of anti-aging. *J. Dermatol. Sci. Cosmet. Technol.* **2024**, *1*, 100014. [CrossRef]
50. Yayla, S.; Hurkul, M.M.; Cetinkaya, A.; Uzun, L.; Ozkan, S.A. Selective apigenin assay in plant extracts and herbal supplement with molecularly imprinted polymer-based electrochemical sensor. *Talanta* **2025**, *281*, 126895. [CrossRef]
51. Bajaj, K.; Adhikary, T.; Gill, P.P.S.; Kumar, A. Edible coatings enriched with plant-based extracts preserve postharvest quality of fruits: A review. *Prog. Org. Coat.* **2023**, *182*, 107669. [CrossRef]
52. Gupta, D.; Lall, A.; Kumar, S.; Patil, T.D.; Gaikwad, K.K. Plant-based edible films and coatings for food-packaging applications: Recent advances, applications, and trends. *Sustain. Food Technol.* **2024**, *2*, 1428–1455. [CrossRef]
53. Sapna; Sharma, C.; Pathak, P.; Yadav, S.P.; Gautam, S. Potential of emerging "all-natural" edible coatings to prevent post-harvest losses of vegetables and fruits for sustainable agriculture. *Prog. Org. Coat.* **2024**, *193*, 108537. [CrossRef]
54. Bodana, V.; Swer, T.L.; Kumar, N.; Singh, A.; Samtiya, M.; Sari, T.P.; Babar, O.A. Development and characterization of pomegranate peel extract-functionalized jackfruit seed starch-based edible films and coatings for prolonging the shelf life of white grapes. *Int. J. Biol. Macromol.* **2024**, *254*, 127234. [CrossRef] [PubMed]
55. Assocle, M.S.; Okidi, L.; Ongeng, D. The effect of rosemary, ginger, or garlic on microbial shelf life and sensory acceptability of nutritionally enriched cassava-based pancake (Kabalagala). *Appl. Food Res.* **2024**, *4*, 100384. [CrossRef]
56. Zaidi, M.; Akbar, A.; Ali, S.; Akram, H.; Ercisli, S.; Ilhan, G.; Sakar, E.; Marc, R.A.; Sonmez, D.A.; Ullah, R.; et al. Application of Plant-Based Edible Coatings and Extracts Influences the Postharvest Quality and Shelf Life Potential of "Surahi" Guava Fruits. *ACS Omega* **2023**, *8*, 19523–19531. [CrossRef] [PubMed]
57. Botelho, L.N.S.; Rocha, D.A.; Braga, M.A.; Silva, A.; de Abreu, C.M.P. Quality of guava cv. 'Pedro Sato' treated with cassava starch and cinnamon essential oil. *Sci. Hortic.* **2016**, *209*, 214–220. [CrossRef]
58. Thakur, R.; Pristijono, P.; Golding, J.B.; Stathopoulos, C.E.; Scarlett, C.J.; Bowyer, M.; Singh, S.P.; Vuong, Q.V. Development and application of rice starch based edible coating to improve the postharvest storage potential and quality of plum fruit (*Prunus salicina*). *Sci. Hortic.* **2018**, *237*, 59–66. [CrossRef]
59. Zhang, X.; Lian, H.; Shi, J.; Meng, W.; Peng, Y. Plant extracts such as pine nut shell, peanut shell and jujube leaf improved the antioxidant ability and gas permeability of chitosan films. *Int. J. Biol. Macromol.* **2020**, *148*, 1242–1250. [CrossRef] [PubMed]
60. Shahid-ul-Islam; Jaiswal, V.; Butola, B.S.; Majumdar, A. Production of PVA-chitosan films using green synthesized ZnO NPs enriched with dragon fruit extract envisaging food packaging applications. *Int. J. Biol. Macromol.* **2023**, *252*, 126457. [CrossRef] [PubMed]
61. Wang, S.; Marcone, M.; Barbut, S.; Lim, L.T. The Impact of Anthocyanin-Rich Red Raspberry Extract (ARRE) on the Properties of Edible Soy Protein Isolate (SPI) Films. *J. Food Sci.* **2012**, *77*, C497–C505. [CrossRef]
62. Kanatt, S.R.; Rao, M.S.; Chawla, S.P.; Sharma, A. Active chitosan-polyvinyl alcohol films with natural extracts. *Food Hydrocoll.* **2012**, *29*, 290–297. [CrossRef]
63. Kumari, M.; Mahajan, H.; Joshi, R.; Gupta, M. Development and structural characterization of edible films for improving fruit quality. *Food Packag. Shelf Life* **2017**, *12*, 42–50. [CrossRef]

64. Coman, M.M.; Oancea, A.M.; Verdenelli, M.C.; Cecchini, C.; Bahrim, G.E.; Orpianesi, C.; Cresci, A.; Silvi, S. Polyphenol content and in vitro evaluation of antioxidant, antimicrobial and prebiotic properties of red fruit extracts. *Eur. Food Res. Technol.* **2018**, *244*, 735–745. [CrossRef]
65. Eça, K.S.; Machado, M.T.C.; Hubinger, M.D.; Menegalli, F.C. Development of Active Films From Pectin and Fruit Extracts: Light Protection, Antioxidant Capacity, and Compounds Stability. *J. Food Sci.* **2015**, *80*, C2389–C2396. [CrossRef] [PubMed]
66. Hu, D.; Liu, X.; Qin, Y.; Yan, J.; Li, J.; Yang, Q. A novel edible packaging film based on chitosan incorporated with persimmon peel extract for the postharvest preservation of banana. *Food Qual. Saf.* **2022**, *6*, fyac028. [CrossRef]
67. Rather, S.A.; Hussain, P.R.; Suradkar, P. Evaluating the effects of chitosan incorporated with seabuckthorn leaf extract composite edible coatings on the shelf life of peach (*Prunus persica* L.) fruits. *Food Humanit.* **2024**, *2*, 100280. [CrossRef]
68. do Nascimento, A.; Toneto, L.C.; Lepaus, B.M.; Valiati, B.S.; Faria-Silva, L.; de São José, J.F.B. Effect of Edible Coatings of Cassava Starch Incorporated with Clove and Cinnamon Essential Oils on the Shelf Life of Papaya. *Membranes* **2023**, *13*, 772. [CrossRef] [PubMed]
69. Sarkar, P.; Dey, A. Phycoremediation—An emerging technique for dye abatement: An overview. *Process Saf. Environ. Prot.* **2021**, *147*, 214–225. [CrossRef]
70. Abera, B.; Duraisamy, R.; Birhanu, T. Study on the preparation and use of edible coating of fish scale chitosan and glycerol blended banana pseudostem starch for the preservation of apples, mangoes, and strawberries. *J. Agric. Food Res.* **2024**, *15*, 100916. [CrossRef]
71. Jost, V.; Kobsik, K.; Schmid, M.; Noller, K. Influence of plasticiser on the barrier, mechanical and grease resistance properties of alginate cast films. *Carbohydr. Polym.* **2014**, *110*, 309–319. [CrossRef] [PubMed]
72. Khan, O.A.; Zaidi, S.; Islam, R.U.; Naseem, S.; Junaid, P.M. Enhanced shelf-life of peach fruit in alginate based edible coating loaded with TiO_2 nanoparticles. *Prog. Org. Coat.* **2023**, *182*, 107688. [CrossRef]
73. Rout, S.S.; Pradhan, K.C. A review on antimicrobial nano-based edible packaging: Sustainable applications and emerging trends in food industry. *Food Control* **2024**, *163*, 110470. [CrossRef]
74. Tran, T.H.; Le, X.C.; Tran, T.N.M.; Nguyen, N.T.T.; Pham, B.N.; Vu, D. Nano selenium–alginate edible coating extends hydroponic strawberry shelf life and provides selenium fortification as a micro-nutrient. *Food Biosci.* **2023**, *53*, 102597. [CrossRef]
75. Nkede, F.N.; Wardak, M.H.; Wardana, A.A.; Fanze, M.; Yan, X.; Jothi, J.S.; Thi Hang Phuong, N.; Tanaka, F.; Tanaka, F. Preparation and characterization of edible coating and film composed of sodium alginate/ylang-ylang oil/cellulose nanocrystals Pickering emulsion and its application to post-harvest control of mandarin (*Citrus reticulata*). *Colloids Surfaces A Physicochem. Eng. Asp.* **2024**, *691*, 133859. [CrossRef]
76. Megha, M.; Gill, P.P.S.; Jawandha, S.K.; Kaur, P.; Sinha, A. Sodium alginate enriched with pomegranate peel extract maintains pear cv. Punjab Beauty cell membrane integrity and postharvest life. *S. Afr. J. Bot.* **2023**, *158*, 149–157. [CrossRef]
77. Zhang, Y.; Shen, Y.; Chen, C.; Gan, Z.; Chen, J.; Wan, C. Loquat leaf extract and alginate based green composite edible coating for preserving the postharvest quality of Nanfeng tangerines. *Sustain. Chem. Pharm.* **2022**, *27*, 100674. [CrossRef]
78. Rojas-Graü, M.A.; Tapia, M.S.; Martín-Belloso, O. Using polysaccharide-based edible coatings to maintain quality of fresh-cut Fuji apples. *LWT* **2008**, *41*, 139–147. [CrossRef]
79. Rojas-Graü, M.A.; Raybaudi-Massilia, R.M.; Soliva-Fortuny, R.C.; Avena-Bustillos, R.J.; McHugh, T.H.; Martín-Belloso, O. Apple puree-alginate edible coating as carrier of antimicrobial agents to prolong shelf-life of fresh-cut apples. *Postharvest Biol. Technol.* **2007**, *45*, 254–264. [CrossRef]
80. Moldão-Martins, M.; Beirão-da-Costa, S.M.; Beirão-da-Costa, M.L. The effects of edible coatings on postharvest quality of the "Bravo de Esmolfe" apple. *Eur. Food Res. Technol.* **2003**, *217*, 325–328. [CrossRef]
81. Rößle, C.; Brunton, N.; Gormley, R.T.; Wouters, R.; Butler, F. Alginate Coating as Carrier of Oligofructose and Inulin and to Maintain the Quality of Fresh-Cut Apples. *J. Food Sci.* **2011**, *76*, H19–H29. [CrossRef] [PubMed]
82. Sarengaowa; Hu, W.; Jiang, A.; Xiu, Z.; Feng, K. Effect of thyme oil–alginate-based coating on quality and microbial safety of fresh-cut apples. *J. Sci. Food Agric.* **2018**, *98*, 2302–2311. [CrossRef] [PubMed]
83. Azarakhsh, N.; Osman, A.; Ghazali, H.M.; Tan, C.P.; Mohd Adzahan, N. Lemongrass essential oil incorporated into alginate-based edible coating for shelf-life extension and quality retention of fresh-cut pineapple. *Postharvest Biol. Technol.* **2014**, *88*, 1–7. [CrossRef]
84. Chen, J.; Wu, A.; Yang, M.; Ge, Y.; Pristijono, P.; Li, J.; Xu, B.; Mi, H. Characterization of sodium alginate-based films incorporated with thymol for fresh-cut apple packaging. *Food Control* **2021**, *126*, 108063. [CrossRef]
85. Tabassum, N.; Khan, M.A. Modified atmosphere packaging of fresh-cut papaya using alginate based edible coating: Quality evaluation and shelf life study. *Sci. Hortic.* **2020**, *259*, 108853. [CrossRef]
86. Zhang, Y.; Pu, Y.; Jiang, H.; Chen, L.; Shen, C.; Zhang, W.; Cao, J.; Jiang, W. Improved sustained-release properties of ginger essential oil in a Pickering emulsion system incorporated in sodium alginate film and delayed postharvest senescence of mango fruits. *Food Chem.* **2024**, *435*, 137534. [CrossRef]

87. Deng, P.; Zhang, Y.; Deng, Q.; Sun, Y.; Li, Y.; Wang, Z.; Jiang, F. Multifunctional sodium alginate-based self-healing edible cross-linked coating for banana preservation. *Food Hydrocoll.* **2024**, *151*, 109753. [CrossRef]
88. Vasconcelos, L.; de Souza, M.; de Oliveira, J.; Filho, E.S.; Silva, A.; Mazzetto, S.E.; Pereira, E.S.; Oliveira, R.L.; Bezerra, L. Elaboration and characterization of bioactive films obtained from the incorporation of cashew nut shell liquid into a matrix of sodium alginate. *Antioxidants* **2021**, *10*, 1378. [CrossRef] [PubMed]

www.ingramcontent.com/pod-product-compliance
Lightning Source LLC
Chambersburg PA
CBHW062045220926
51850CB00010B/2672

MDPI AG
Grosspeteranlage 5
4052 Basel
Switzerland
Tel.: +41 61 683 77 34

Colloids and Interfaces Editorial Office
E-mail: colloids@mdpi.com
www.mdpi.com/journal/colloids